Lecture Notes in Computer Science 6968

Commenced Publication in 1973
Founding and Former Series Editors:
Gerhard Goos, Juris Hartmanis, and Jan van Leeuwen

Elena A. Troubitsyna (Ed.)

Software Engineering for Resilient Systems

Third International Workshop, SERENE 2011
Geneva, Switzerland, September 29-30, 2011
Proceedings

 Springer

Volume Editor

Elena A. Troubitsyna
Åbo Akademi University, Department of IT
Joukahaisenkatu 3-5A, 20520 Turku, Finland
E-mail: elena.troubitsyna@abo.fi

ISSN 0302-9743   e-ISSN 1611-3349
ISBN 978-3-642-24123-9   e-ISBN 978-3-642-24124-6
DOI 10.1007/978-3-642-24124-6
Springer Heidelberg Dordrecht London New York

Library of Congress Control Number: 2011936358

CR Subject Classification (1998): C.2.4, D.2, D.4.5, H.4, F.3, D.3

LNCS Sublibrary: SL 2 – Programming and Software Engineering

Typesetting: Camera-ready by author, data conversion by Scientific Publishing Services, Chennai, India

Printed on acid-free paper

Springer is part of Springer Science+Business Media (www.springer.com)

Preface

This volume contains the proceedings of the Third International Workshop on Software Engineering for Resilient Systems (SERENE 2011). SERENE 2011 took place in Geneva, Switzerland, during September 29–30, 2011. The SERENE workshop is an annual event that brings together researchers and practitioners working on the various aspects of design, verification and assessment of resilient systems. In particular it covers such areas as:

- Modelling of resilience properties: formal and semi-formal techniques
- Requirements, software engineering and re-engineering for resilience
- Verification and validation of resilient systems
- Resilience prediction and experimental measurement
- Error, fault and exception handling in the software life-cycle
- Frameworks, patterns and software architectures for resilience
- Resilience at run-time: metadata, mechanisms, reasoning and adaptation
- Engineering of self-healing autonomic systems
- Quantitative approaches to ensuring resilience
- CASE tools for developing resilient systems

SERENE 2011 featured two invited speakers – Ivica Crnkovic and Peter Popov. Ivica Crnkovic, Mälardalen University, Sweden, is an expert in component-based design of dependable systems. He is a leader of the PROGRESS Centre for Predictable Embedded Software Systems—the center that conducts world-class research on advanced methods and tools for engineering dependable software-intensive systems. Peter Popov, City University of London, is well known for his work on software reliability assessment, software safety, fault-tolerance and more recently critical infrastructure protection. His work on preliminary interdependency analysis constitutes cutting-edge research in modelling and assessment of complex critical infrastructures.

The workshop was established by the members of the ERCIM working group SERENE. The group promotes the idea of resilient-explicit development process. It stresses the importance of extending the traditional software engineering practice with theories and tools supporting modelling and verification of various aspects of resilience. We would like to thank the SERENE working group for their hard work in publicizing the event and contributing to its technical program.

SERENE 2011 attracted 20 submissions, and accepted 13 papers. All papers received four rigorous reviews. The large number of high-quality submissions allowed us to build a strong and technically enlightening program. We would like to express our gratitude to the Program Committee members and the additional reviewers who actively participated in reviewing and discussing the submissions.

Organization of such a workshop is challenging. We would like to acknowledge the help of the technical and administrative staff of Newcastle University and the University of Geneva. SERENE 2011 was supported by ERCIM (European Research Consortium in Informatics and Mathematics), FP7 DEPLOY IP (on Industrial deployment of system engineering methods providing high dependability and productivity), University of Geneva and LASSY (Laboratory for Advanced Software Systems, University of Luxembourg).

July 2011   Elena Troubitsyna
                                                                Program Chair

                                                                 Didier Buchs
                                                                General Chair

Organization

Steering Committee

Nicolas Guelfi   University of Luxembourg
Henry Muccini   University of L'Aquila, Italy
Patrizio Pelliccione   University of L'Aquila, Italy
Alexander Romanovsky   Newcastle University, UK

General Chair

Didier Buchs   University of Geneva, Switzerland

Program Chair

Elena Troubitsyna   Åbo Akademi University, Finland

Program Committee

Finn Arve Aagesen	NTNU, Norway
Paris Avgeriou	University of Groningen, The Netherlands
Antonia Bertolino	CNR-ISTI, Italy
Andrea Bondavalli	University of Florence, Italy
Felicita Di Giandomenico	CNR-ISTI, Italy
Giovanna Di Marzo Serugendo	University of Geneva, Switzerland
John Fitzgerald	Newcastle University, UK
Stefania Gnesi	CNR-ISTI, Italy
Vincenzo Grassi	University of Rome Tor Vergata, Italy
Nicolas Guelfi	University of Luxembourg
Paola Inverardi	University of L'Aquila, Italy
Valerie Issarny	INRIA, France
Mohamed Kaaniche	LAAS-CNRS, France
Vyacheslav Kharchenko	National Aerospace University, Ukraine
Jörg Kienzle	McGill University, Canada
Paul Klint	CWI, The Netherlands
Linas Laibinis	Åbo Akademi University, Finland
Tom Maibaum	McMaster University, Canada
Raffaela Mirandola	Politecnico di Milano, Italy
Ivan Mistrik	Indep. Consultant, Germany
Henry Muccini	University of L'Aquila, Italy
Andras Pataricza	BUTE, Hungary
Patrizio Pelliccione	University of L'Aquila, Italy

Ernesto Pimentel Sanchez	SpaRCIM, Spain
Alexander Romanovsky	Newcastle University, UK
Anthony Savidis	FORTH, Greece
Peter Schneider-Kamp	University of Southern Denmark
Vidar Slåtten	NTNU, Norway
Francis Tam	Nokia, Finland
Apostolos Zarras	University of Ioannina, Greece

Reviewers

Dan Tofan	University of Groningen, The Netherlands
Uwe Van Heesch	University of Groningen, The Netherlands
Alban Linard	University of Geneva, Switzerland
Matteo Risoldi	University of Geneva, Switzerland
Mieke Massink	CNR-ISTI, Italy

Table of Contents

Requirements Engineering and Product Lines

Invited Talk

Monitoring and Self-adaptation

Security and Intrusion Avoidance

Preliminary Interdependency Analysis (PIA): Method and Tool Support

Peter Popov

Centre for Software Reliability, City University London,
Northampton Square, London, EC1V 0HB, UK
ptp@csr.city.ac.uk

Abstract. One of the greatest challenges in enhancing the protection of Critical Infrastructures (CIs) against accidents, natural disasters, and acts of terrorism is establishing and maintaining an understanding of the interdependencies between infrastructures. Understanding interdependencies is a challenge both for governments and for infrastructure owners/operators. Both, to a different extent, have an interest in services and tools that can enhance their risk assessment and management to mitigate large failures that may propagate across infrastructures. The abstract presents an approach (the method and tool support) to interdependency analysis developed recently by Centre for Software Reliability, City University London. The method progresses from a qualitative phase during which a fairly abstract model is built of interacting infrastructures. Via steps of incremental refinement more detailed models are built, which allow for *quantifying* interdependencies. The tool support follows the methodology and allows analysts to build *quickly* models at the appropriate level of abstraction (qualitative or very detailed including deterministic models specific for the particular domain, e.g. various flow models). The approach was successfully applied to a large scale case-study with more than 800 modeling elements.

Keywords: critical infrastructures, interdependencies, stochastic modeling.

1 Introduction

One of the greatest challenges in enhancing the protection of Critical Infrastructures (CIs) against accidents, natural disasters, and acts of terrorism is establishing and maintaining an understanding of the interdependencies between infrastructures and the dynamic nature of these interdependencies. Interdependency can be a source of "unforeseen" threat when failure in one infrastructure may cascade to other infrastructures, or it may be a source of resilience in times of crisis; e.g., by re-allocating resources from one infrastructure to another.

Understanding interdependencies is a challenge both for governments and for infrastructure owners/operators. Both, to a different extent, have an interest in services and tools that can enhance their risk assessment and management to mitigate large failures that may propagate across infrastructures. However, cost of investment in infrastructure modelling and interdependency analysis tools and methods, including

E.A. Troubitsyna (Ed.): SERENE 2011, LNCS 6968, pp. 1–8, 2011.

the supporting technology, may reach millions of pounds, depending on the size of the system to be modelled, on the level of detail and on the mode of modelling (real-time or off-line). These factors will determine the software, hardware, data and personnel requirements.

It is therefore very important to understand what the scope and the overall requirements of an interdependency analysis service are going to be, before proceeding with such an investment. However, the decision on what modelling and visualisation capabilities are needed is far from simple. Detailed requirements may not be understood until some modelling and simulation has been conducted already, in order to identify critical dependencies and decide what level of fidelity is required to investigate them further.

This abstract presents an approach to interdependency analysis that attempts to address these challenges; the approach—*Preliminary Interdependency Analysis (PIA)*—starts off at a high-level of abstraction, supporting a cyclic, systematic thought process that can direct the analysis towards identifying lower-level dependencies between components of CIs. Dependencies can then be analysed with probabilistic models, which would allow one to conduct studies focussed on identifying different measures of interests, e.g. to establish the likelihood of cascade failure for a given set of assumptions, the weakest link in the modelled system, etc. If a high-fidelity analysis is required, PIA can assist in making an informed decision of what to model in more detail. The method is applicable as both:

- a lightweight method and accessible to Small-to-Medium Enterprises (SMEs) in support of their business continuity planning (e.g., to model information infrastructure dependencies, or dependencies on external services such as postal services, couriers, and subcontractors);
- a heavyweight method of studying with an increasing level of detail the complex regional and nationwide CIs combining probabilistic and deterministic models of CIs.

PIA is supported by a toolkit; the *PIA Toolkit* is based on two 3rd party software applications:

The **PIA Designer**, which allows a modeller to define a model of interdependent CIs and define the parameters needed for any quantitative study. For visual representation the tool uses a proprietary tool *Asce* (http://www.csr.city.ac.uk/projects/cetifs.html).

The **Execution engine**, which allows for executing a model developed with the PIA Designer, i.e. a simulation study based on the model to be conducted and the measures of interest to be collected. The Execution engine uses *Möbius* (http://www.mobius.illinois.edu/), customised extensively with a bespoke proprietary development.

The current version of the toolkit allows for two main categories of models:

- Model of interdependent CIs at a fairly high level of abstraction (i.e. without detailed modelling of the networks used by the respective services).
- The model can be parameterised and then the simulation executable can be deployed on the Execution Engine.

- As above but adding any degree of detail that the modeller may consider necessary including high fidelity deterministic models available as 3rd party software modules

The method supported by the toolkit was successfully applied to a range of case studies – from a relatively simple IT infrastructure of an SME (a couple of dozens of modelled elements) to a regional system of two CIs namely the power grid and telecommunication network around Rome, Italy (with 800+ modelled elements).

2 Method: Preliminary Interdependency Analysis (PIA)

Preliminary Interdependency Analysis (PIA) is an analysis activity that seeks to understand the range of possible interdependencies and provide a justified basis for further modelling and analysis. Given a collection of CIs, the objectives of PIA are to develop, through a continuous, cyclical process of refinement, an appropriate *service model* for the infrastructures, and to document assumptions about resources, environmental impact, threats and other factors.

PIA has several benefits; in particular, PIA can:
- help one to discover and better understand dependencies which may be considered as "obvious" and as such are often overlooked (e.g. telecommunications need power)
- support the need for agile and time-efficient analyses (cannot always wait for the high fidelity simulation)
- be also used by Small and Medium Enterprises (SMEs) and not just infrastructure owners and government

PIA allows for the creation and refinement of interdependency models, in a focused manner, by revisiting earlier stages in the PIA process in the light of the outcomes of latter stages. For example, an initial application of PIA should result in a sufficiently concrete and clearly defined model of CIs (and their dependencies). However, following the first design iteration, an analysis of the model could cause us to question the assumptions made earlier on in the design process. As a consequence, the model may be revised and refined; as we shall see later on, revisiting previous phases of the development process is a key aspect of the PIA method and philosophy overall. PIA is broadly broken up in two parts:

Qualitative analysis. The modelling exercise begins with a definition of the boundaries of the system to be studied and its components. Starting off at a high level, the analyst may go through a cyclical process of definitions, but may also be focused on a particular service, so the level of detail may vary between the different parts of the overall model. The identification of dependencies (service-based or geographical) will start at this point.

Quantitative analysis. The models created during the qualitative PIA are now used to construct an *executable*, i.e. a simulator of the model behaviour in the presence of failures of the modelled entities for the chosen model parameterisation. The model parameterisation may be based either on expert judgement or on analysis of incident data. Examples of such data analyses and fitting the available data to plausible probabilistic data models were developed recently [1].

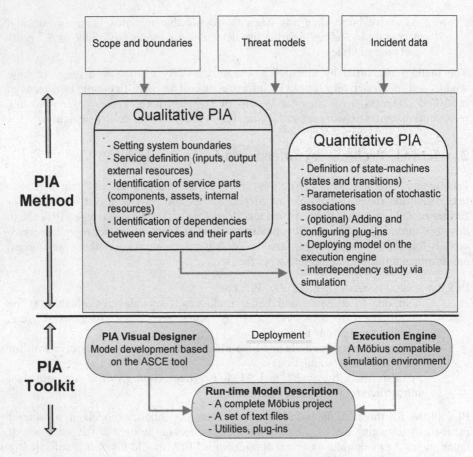

Fig. 1. Overview of PIA method and toolkit

The PIA Toolkit provides support for both the qualitative and quantitative analyses. Figure 1 illustrates an overview of the method and the toolkit.

The interdependency models, of course, have to be related to a purpose and this should be captured in terms of a scenario and related requirements. The narrative aspect of the scenario is enormously important as it provides the basis for asking questions and discovering interdependencies as the starting point for more formal models.

Typically the systems of interdependent CIs of interest are complex: include many services which in turn consist of many parts. Given the complexity and size of the analysed systems tool support is essential.

3   PIA Model Architecture: Two Levels of Abstraction

PIA models broadly operate at two distinct levels of abstraction.

Model of interacting services (service-level model). The modelled CIs are represented by a set of interdependent services. Here, the view is purposefully abstract, so that we can reason about dependencies among the services (i.e. data centre X depends on power plant Y); Service-level dependencies are elicited by the defined lower-level dependencies among each service's constituent entities (physical components, resources etc.). These associations among components are referred to within PIA as *coupling points*. The coupling points *incoming* to a service can be associated with the resources that the service requires (e.g., a telecommunication service consumes "commodities" supplied by a power service). The resources consumed by a service can be obtained from the organisation's reserves *(internal resources)* or provided by another organisation *(external resources)*. The *outgoing* coupling points instead define how the outputs from a service get consumed by other services (as either inputs or resources).

Detailed service behaviour model (DSBM). Implementation details are provided for an *individual service,* e.g. the networks upon which a particular service relies. For instance a GSM telecommunication operator typically relies on a network of devices deployed to cover a particular area (e.g. masts, etc.). Via DSBM we can choose the level of detail used to model these networks. In the example above DSBM may range from a connectivity graph – which cells of the network are connected with each other to a high fidelity model of the protocols used in the GSM network. We tend to think of DSBM as the networks owned (at least partially and/or maintained) by the respective service operator, i.e. an organisation. Although such a view is not necessary, it allows one to model via DSBM several important aspects. For instance the level of investment and the culture (strong emphasis on engineering vs. outsourcing the maintenance) within the organisation will affect how well the network is maintained (i.e. frequency of outages and speed of recovery). Thus, the process of recovery (a parameter used in DSBM) can be a useful proxy of the level of investment. Thus, through DSBM one can study scenarios which at first may seem outside the scope of PIA. An example of such a scenario would be comparing the deregulation vs. tight regulation in critical CIs.

4 PIA Stages

PIA is carried out in seven stages (Figure 2):

4.1 Stage 1

CI description and scenario context. A CI description provides a concrete context and concept of operation. This is the first level of scoping for the analysis task; the CI description gives the first indications of analysis boundaries. DSBM entities are identified and recorded.

4.2 Stage 2

Model development. A model of the services (resources, inputs, outputs, system's states) and the operational environment and system boundaries are developed, based on the CI description. Model boundary definitions are used at this stage to further restrict the scope of the analysis. Dependencies between the services are identified and the *coupling points* are defined: these refer on the one hand to the inputs and resources required by each of the services and one the other hand – to the outputs that each of the services produces.

4.3 Stage 3

DSBM model development. DSBMs are defined by selecting the right level of abstraction for the services: some of the services may be treated as black-boxes; in this case their representation in the DSBM will require no refinement in comparison with Stage 2. For those services, which are modelled in more detail one starts by defining explicitly their components and the assets including resort to using *existing models* of the underlying *physical* networks used by the services. A level of consistency is achieved between the service model and DSBM: the coupling points appear in both Views.

4.4 Stage 4

Initial dependency and interdependency identification. While some of the service dependencies have already been identified and recorded in Stage 2 (via input/output/resource identification), at this stage the modeller looks for additional sources of dependence (e.g. common components/assets), which may make several services vulnerable to common faults or threats. These can be derived by examining the service-level model, taking into account other contextual information (e.g. scenarios, threat models, attacker profile). The captured dependencies are modelled as *stochastic association* between the services or components thereof. Each stochastic association is seen as a relationship between a parent and a child: the state of the parent affects the modelled behaviour of the child.

4.5 Stage 5

Probabilistic model development. Since we are dealing with risk, we take the view that, given the state space formed by the modelled entities (MEs), a stochastic process must be constructed upon it that captures the unpredictable nature of the states of the MEs, their changes and the interactions between CIs over time. In this stage probabilistic models of the MEs are defined. These are *state-machines*, a well known formalism in software engineering, modelled after the formalism used in the *Stochastic Activity Networks (SANs)*.

4.6 Stage 6

(optional) Adding deterministic models of behaviour. At this stage the modeller may decide to extend the behaviour of the probabilistic model adding deterministic models of behaviour. Such a step may be useful when the modeller is seeking to extend the fidelity of the simulation beyond the standard mechanisms possible with a pure probabilistic model.

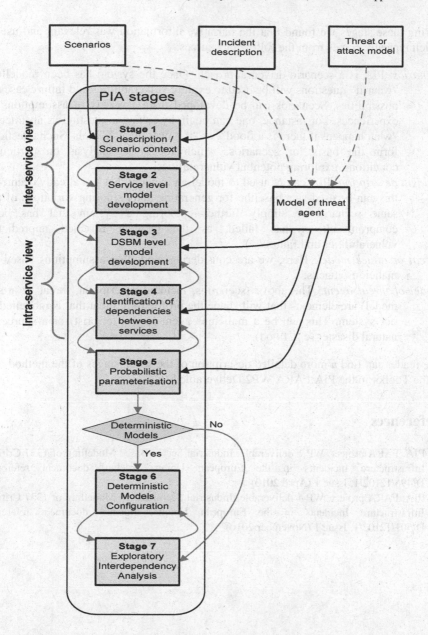

Fig. 2. PIA method stages

4.7   Stage 7

Exploratory interdependency analysis. A Monte Carlo simulation is used to quantify the impact of interdependencies on the behaviour of the system under study and draw more conclusions about the probability of interdependency-related risk.

During these stages we found that the narrative information was relevant and useful which typically comes from the following sources:

Scenarios: PIA is a scenario-driven approach. Once the system has been modelled, "what-if" questions will be used to explore vulnerabilities and failure cascade possibilities. Scenarios can be developed from a variety of assumptions or experiences. For instance, one can begin by asking a question as abstract as "what happens if there is a flood", or "if power plant X" fails. Such questions form the basis for scenarios, which focus the analysis on particular conditions, exploring potential vulnerabilities.

Incident description: PIA can be used to model an incident that has already occurred; this can be used as a baseline for generating and exploring variations of the same scenario or simply further exploring a system that has been compromised, or has failed, as the incident revealed unpredicted vulnerabilities and failures.

Threat or attack model: Here, we are considering modelling assumptions based on malicious attacks.

Model of threat agent: The above (scenarios, incident description, threat or attack model) are elements that will shape the profile of a threat that is modelled in our system. This can be a malicious agent (e.g. a terrorist) or a source of natural disaster (e.g. flood).

The reader can find a more detailed description of the seven stages of the method and of the Toolkit in the PIA:FARA WP2 Deliverable [2].

References

1. PIA: FARA project, WP 1 deliverable, Industrial Sector-Based Modelling of 1337 Critical Infrastructure Incidents in the European Union, Adelard document reference D/496/12102/1, Issue 1 (April 2010)
2. PIA: FARA project, WP 1 deliverable, Industrial Sector-Based Modelling of 1337 Critical Infrastructure Incidents in the European Union", Adelard document reference D/501/12102/1, Issue 1 (November 2010)

Use Case Scenarios as Verification Conditions: Event-B/Flow Approach

Alexei Iliasov

Newcastle University, UK
alexei.iliasov@ncl.ac.uk

Abstract. Model-oriented formalisms rely on a combination of safety constraints and satisfaction of refinement obligations to demonstrate model correctness. We argue that for a significant class of models a substantial part of the desired model behaviour would not be covered by such correctness conditions, meaning that a formal development potentially ends with a correct model inadequate for its purpose. In this paper we present a method for augmenting Event-B specifications with additional proof obligations expressed in a visual, diagrammatic way. A case study illustrates how the method may be used to strengthen a model by translating use case scenarios from requirement documents into formal statements over a modelled system.

1 Introduction

Use cases are a popular technique for the validation of software systems and constitute an important part of requirements engineering process. It is an essential part of the description of functional requirements of a system. There exists a vast number of notations and methods supporting the integration of use cases in a development process (see [5] for a structured survey of use case notations). With few exceptions, the overall aim is the derivation of test inputs for the testing of the final product. Our approach is different. We propose to exploit use cases in the course of a step-wise formal development process for the engineering of correct-by-construction systems. It is assumed that a library of use case scenarios is available together with a system requirements document and use cases are presented in a sufficiently precise manner. We discuss a technique and tool for expressing use case scenarios as formal verification conditions that become a part of a formal model of a developed system. It is guaranteed that the final product obtained from such a model posesses, by the virtue of the development method, all the properties expressed in use cases scenarios. The overall approach is generally in line with some existing work on formalisation and formal validation of use cases [7]. The approach is investigated on the basis of an extension of the Event-B modelling method [1] with a technique for formally capturing use case scenarios as theorems over model state.

The paper is organised as follows. Section 2 gives a brief overview of Event-B notation and methodology. The motivation behind the approach is presented in Section 3. The essential details of the approach and its integration with Event-B

E.A. Troubitsyna (Ed.): SERENE 2011, LNCS 6968, pp. 9–23, 2011.

```
MACHINE M
  SEES Context
  VARIABLES v
  INVARIANT I(c, s, v)
  INITIALISATION ...
  EVENTS
    E₁ = any vl where g(c, s, vl, v) then S(c, s, vl, v, v') end
    ...
END
```

Fig. 1. Event-B model structure

are presented in Section 4. Section 5 gives examples of use cases in the role of validation conditions for an Event-B model of a networking file system.

2   Event-B

Event-B [1] is a formal modelling method for the realisation of correct-by-construction software systems.

An Event-B development starts with the creation of a very abstract specification. The cornerstone of Event-B method is the stepwise development that facilitates a gradual design of a system implementation through a number of correctness-preserving *refinement* steps. The general form of an Event-B model (or *machine*) is shown in Figure 1. Such a model encapsulates a local state (program variables) and provides operations on the state. The actions (called *events*) are defined by a list of new local variables (parameters) vl, a state predicate g called *event guard*, and a next-state relation S called *substitution* (see the **EVENTS** section in Figure 1).

The **INVARIANT** clause contains the properties of the system (expressed as state predicates) that should be preserved during system execution. These define *safe states* of a system. In order for a model to be consistent, invariant preservation should be formally demonstrated. Data types, constants and relevant axioms are defined in a separate component called *context*.

Model correctness is demonstrated by generating and discharging a collection of proof obligations. There are proof obligation demonstrating model consistency, such the preservation of the invariant by the events, and the refinement link to another Event-B model. Putting it as a requirement that an enabled event produces a new state v' satisfying the model invariant, the model *consistency* condition states that whenever an event on an initialisation action is attempted, there exists a suitable new state v' such that the model invariant is maintained - $I(v')$. This is usually stated as two separate conditions: event feasibility and event invariant satisfaction.

FIS $I(c, s, v) \wedge g(c, s, vl, v) \Rightarrow \exists v'. \ S(c, s, vl, v, v')$
INV $I(c, s, v) \wedge g(c, s, vl, v) \wedge S(c, s, vl, v, v') \Rightarrow I(c, s, v')$

The consistency of Event-B models, i.e., verification of well-formedness and invariant preservation as well as correctness of refinement steps is demonstrated by discharging relevant *proof obligations* (such as **INV** and **FIS**) that, collectively, define the *proof semantics* of a model. The Rodin platform [10], a tool supporting Event-B, is an integrated environment that automatically generates necessary proof obligations and manages a collection of automated provers and solvers autonomously discharging the generated theorems.

3 Motivation

A stack may be defined in Event-B as a pair of variables $stack \in 1 \mathinner{.\,.} top \to \mathbb{N}$ and $top \in \mathbb{N}$ where top is the stack top pointer and $stack$ is a sequence of values representing a stack contents. The following two events define the obvious stack manipulation operations.

push = **any** v **where** $v \in \mathbb{N}$ **then** $stack(top + 1) := v \parallel top := top + 1$ **end**
pop = **when** $top > 0$ **then** $stack := 1 \mathinner{.\,.} top - 1 \lhd stack \parallel top := top - 1$ **end**

It is not difficult to ascertain that the above model is indeed a specification of a stack. However, the model invariants (the typing conditions for variables $stack$ and top given above) permit a wide range of safe but undesirable behaviours. For instance, we could have made a mistake in the definition of event *push*

push_broken = **any** v **where** $v \in \mathbb{N}$ **then** $stack(top + 1) := v$ **end**

Event push_broken does not violate the invariants and hence there is no feedback from the verification tools. Although the model is correct it cannot be used in the role of a stack, that is, it is an inadequate representation of stack. The problem is not in the specification itself: the invariant is as strong as it can be for this model. While, in some cases, it is possible to find a model abstraction [1] that, via the refinement relation, would demonstrate the necessary conditions, this is not always possible in practice due to often awkward models and interference with the use of refinement as a development method.

The problem is not artificial: in larger models, it is difficult to identify a problem by simply reading model text. An informal inspection of some recent Event-B developments [3] shows that industrially-inclined models (models of a piece of software rather than models of protocols or algorithms) tend to have less restrictive (but not necessarily less numerous) invariants, more involved event actions and exhibit the prevalence of horizontal refinement - a form of data refinement where all new model elements contribute to the behaviour on the new, hidden state. Correspondingly, in such models, much of behavioural specifications is not under an obligation to establish any verification conditions.

---

[1] One attempt at such an abstraction may be found [8].

The issue of constructing a correct model that does what is expected from it is generally known as a problem of model adequacy and largely falls into the domain of requirements engineering. Here we study an application of one requirements engineering technique - use case scenarios - in the formal setting of the Event-B method. To illustrate the main point of the proposed technique let us consider the following (algebraic) specification of a stack object S holding elements of type $\mathbb{N}$.

$$
\begin{aligned}
&init \in S & &empty(init) = \text{TRUE} \\
&empty : S \to \text{BOOL} & &empty(push(s,v)) = \text{FALSE} \\
&push : S \times \mathbb{N} \to S & &top(push(s,v)) = v \\
&pop : S \to S \times \mathbb{N} & &pop(push(s,v)) = s
\end{aligned}
$$

In a contrast to the Event-B model above, it does not specify how the stack operations update the stack but rather defines few principal properties of stacks. This style makes it easier to spot unexpected model properties as the defining characteristics are given in an explicit and concise form whereas for a model-oriented formalism, like Event-B, one has to do some mental calculations.

An interesting exercise is to translate such algebraic properties into Event-B model theorems (that is, equivalent statements over $stack$ and top). Notice that $pop(push(s,v)) = s$ may be put as

$$(\mathsf{pop} \circ \mathsf{push}) \subseteq \mathrm{id}(\Sigma)$$

where Σ is the state of the model: $\Sigma = \{s \times t \mid t \in \mathbb{N} \wedge s \in 1\,..\,t \to \mathbb{N}\}$; s and t are shorthands for $stack$ and top. The relations $push$ and pop are easily derived from the event definitions:

$\mathsf{pop} = \{(s \mapsto t) \mapsto (s' \mapsto t') \mid t > 0 \wedge s' = 1\,..\,t-1 \triangleleft s \wedge t' = t-1\}$
$\mathsf{push} = \{(s \mapsto t) \mapsto (s' \mapsto t') \mid v \in \mathbb{N} \wedge s' = s \triangleleft \{t+1 \mapsto v\} \wedge t' = t+1\}$

The condition is proven by expanding the relational composition into a join:

$\mathsf{pop} \circ \mathsf{push} = \{(s \mapsto t) \mapsto (s'' \mapsto t'') \mid v \in \mathbb{N} \wedge s' = s \cup \{t+1 \mapsto v\} \wedge t' = t+1 \wedge$
$\qquad\qquad\qquad\qquad\qquad t' > 0 \wedge s'' = 1\,..\,t'-1 \triangleleft s' \wedge t'' = t'-1\}$
$\qquad = \{(s \mapsto t) \mapsto (s'' \mapsto t'') \mid \cdots \wedge s'' = s \wedge t'' = t\}$
$\qquad = \{x \mapsto x \mid x \in \Sigma\} = \mathrm{id}(\Sigma)$

In the same manner, one can establish that condition $\mathsf{pop} \circ \mathsf{push\_broken} \subseteq \mathrm{id}(\Sigma)$ does not hold and therefore formally rule out the $\mathsf{push\_broken}$ version of the $push$ event.

The main difficulty in checking $(\mathsf{pop} \circ \mathsf{push}) \subseteq \mathrm{id}(\Sigma)$ is the construction of the verification goal from the event definitions. The resultant theorem is trivial for an automated prover. The proposal we discuss in this paper allows one to construct this kind of theorems very easily and in large quantities (when necessary) using a simple visual notation.

4   Flow Language

The section presents the semantics of the Flow language and its visual notation.

4.1   Flow theorems

Consider the following relations over pairs of relations on some set S.

$$U = \{f \mapsto g \mid \varnothing \subset f \subseteq S \times S \wedge \varnothing \subset g \subseteq S \times S\}$$
$$\mathbf{ena} = \{f \mapsto g \mid f \mapsto g \in U \wedge \mathrm{ran}(f) \subseteq \mathrm{dom}(g)\}$$
$$\mathbf{dis} = \{f \mapsto g \mid f \mapsto g \in U \wedge \mathrm{ran}(f) \cap \mathrm{dom}(g) = \varnothing\}$$
$$\mathbf{fis} = \{f \mapsto g \mid f \mapsto g \in U \wedge \mathrm{ran}(f) \cap \mathrm{dom}(g) \neq \varnothing\}$$

The definitions are concerned with the properties of a composite relation $g \circ f$. f **ena** g states that $g \circ f$ is defined for every value on which f is defined - $dom(g \circ f) = dom(f)$; relation f **dis** g implies that $g \circ f = \varnothing$; f **fis** g means that there is at least one pair of values satisfying relation $g \circ f$: $\exists v, u \cdot u \, (g \circ f) \, v$. The relations enjoy the following properties.

$$\mathbf{dis} \cap \mathbf{fis} = \varnothing \Leftrightarrow \neg(f \, \mathbf{dis} \, g \wedge f \, \mathbf{fis} \, g)$$
$$\mathbf{ena} \cap \mathbf{dis} = \varnothing \Leftrightarrow \neg(f \, \mathbf{ena} \, g \wedge f \, \mathbf{dis} \, g)$$
$$\mathbf{dis} \cup \mathbf{fis} = U \Leftrightarrow f \, \mathbf{dis} \, g \vee f \, \mathbf{fis} \, g$$

Let p_e, G_e, R_e characterise the parameters, guard and action of an event e. Assuming that the consistency proof obligations are discharged, the universe of system states Σ is said to be its safe states: $\Sigma = \{v \mid I(v)\}$. An event e is a next-state relation of the form $e \subseteq \Sigma \times \Sigma$. Let $\mathsf{before}(e) \subseteq \Sigma$ and $\mathsf{after}(e) \subseteq \Sigma$ signify the domain and the range of relation e. The set $\mathsf{before}(e)$ corresponds to the enabling states defined by the event guard and $\mathsf{after}(e)$ is a set of possible new states computed by the event:

$$\mathsf{before}(e) = \{v \mid I(v) \wedge \exists p_e \cdot G_e(p_e, v)\}$$
$$\mathsf{after}(e) = \{v' \mid I(v) \wedge \exists p_e \cdot (G_e(p_e, v) \wedge R_e(p_e, v, v'))\}$$

Let b and h be some events. Taking into the account the definitions of before and after, relations $\mathbf{ena}, \mathbf{dis}, \mathbf{fis}$ may be expanded as follows.

$$b \, \mathbf{ena} \, h \Leftrightarrow \mathsf{after}(b) \subseteq \mathsf{before}(h)$$
$$\Leftrightarrow \{v' \mid I(v) \wedge \exists p_b \cdot (G_b(p_b, v) \wedge R_b(p_b, v, v'))\}$$
$$\subseteq \{v \mid I(v) \wedge \exists p_h \cdot G_h(p_h, v)\}$$
$$\Leftrightarrow \forall v, v', p_b \cdot I(v) \wedge G_b(p_b, v) \wedge R_b(p_b, v, v') \Rightarrow \exists p_h \cdot G_h(p_h, v') \quad (\mathbf{FENA})$$

$$b \, \mathbf{dis} \, h \Leftrightarrow \mathsf{after}(b) \cap \mathsf{before}(h) = \varnothing$$
$$\Leftrightarrow \mathsf{after}(b) \subseteq \Sigma \setminus \mathsf{before}(h)$$
$$\Leftrightarrow \mathsf{after}(b) \subseteq \{v \mid I(v)\} \setminus \{v \mid I(v) \wedge \exists p_h \cdot G_h(p_h, v)\}$$
$$\Leftrightarrow \mathsf{after}(b) \subseteq \{v \mid I(v) \wedge \forall p_h \cdot \neg G_h(p_h, v)\}$$
$$\Leftrightarrow \forall v, v', p_b, p_h \cdot I(v) \wedge G_b(p_b, v) \wedge R_b(p_b, v, v') \Rightarrow \neg G_h(p_h, v') \quad (\mathbf{FDIS})$$

$$b \, \mathbf{fis} \, h \Leftrightarrow \mathsf{after}(b) \cap \mathsf{before}(h) \neq \varnothing$$
$$\Leftrightarrow \exists v, v', p_b, p_h \cdot I(v) \wedge G_b(p_b, v) \wedge R_b(p_b, v, v') \wedge G_h(p_h, v') \quad (\mathbf{FFIS})$$

Conditions ($\mathbf{FENA}, \mathbf{FDIS}, \mathbf{FFIS}$) are the main flow verification conditions. Event-B proof obligations of these form are automatically derived by a tool supporting the approach.

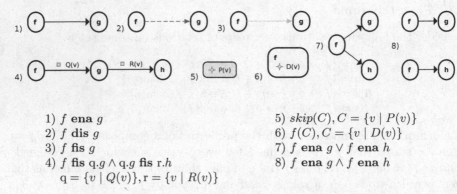

1) f **ena** g
2) f **dis** g
3) f **fis** g
4) f **fis** q.$g \wedge$ q.g **fis** r.h
    q $= \{v \mid Q(v)\}$, r $= \{v \mid R(v)\}$

5) $skip(C), C = \{v \mid P(v)\}$
6) $f(C), C = \{v \mid D(v)\}$
7) f **ena** $g \vee f$ **ena** h
8) f **ena** $g \wedge f$ **ena** h

Fig. 2. A summary of the core flow notation and its interpretation

Assumptions and assertions It is convenient to construct new events from existing events. We define two operators for this: *assumption* and *assertion*. An assumption constraints the enabling set of an event while an assertion constraints the set of new states computed by the event.

$$P.e \stackrel{\text{def}}{=} \{t \mapsto r \mid t \mapsto r \in e \wedge t \in P\} \qquad \mathsf{before}(P.e) = \mathsf{before}(e) \cap P$$
$$e.Q \stackrel{\text{def}}{=} \{t \mapsto r \mid t \mapsto r \in e \wedge r \in Q\} \qquad \mathsf{after}(e.Q) = \mathsf{after}(e) \cap Q$$

It is trivial to see that such constrained events are safe: $P.e \subseteq e \wedge e.Q \subseteq e$. One special case is a constraint used in the roles of both assumption and assertion: $e(D) = D.e \cap D.e$. It is easy to see that $e(D)$ may be obtained by adding a guard predicate L to event e such that $D = \{v \mid L(v)\}$. It often necessary to make statement about a state rather than an event. This is done with the help of the skip event that does not change system state: skip $= \mathrm{id}(\Sigma)$. Then skip(D) is a stuttering step constrained to states D.

4.2   Graphical Notation

The approach is realised by a tool employing a visual, diagrammatic depiction of Flow theorems. A Flow diagram always exists in an association with one Event-B model. The theorems expressed in a Flow diagram are statements about the behaviour of the associated model. The basic element of a diagram is an event, visually depicted as a node (in Figure 2, f and g represent events). Event definition (its parameters, guard and action) is imported from the associated Event-B model. One special case of node is skip event, denoted by a grey node colour (Figure 2, 5). Event relations **ena, dis, fis** are represented by edges connecting nodes ((Figure 2, 1-3)). Depending on how a diagram is drawn, edges (flow theorems) are said to be in *and* or *or* relation (Figure 2, 7-8). New events are derived from model events by strengthening their guards (a case of symmetric assumption and assertion) (Figure 2, 6). Edges may be annotated with constraining predicates inducing assertion and assumption derived events (Figure 2, 4). Not

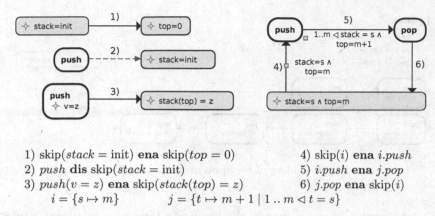

1) skip($stack = $ init) **ena** skip($top = 0$)
2) $push$ **dis** skip($stack = $ init)
3) $push(v = z)$ **ena** skip($stack(top) = z$)
   $i = \{s \mapsto m\}$

4) skip(i) **ena** $i.push$
5) $i.push$ **ena** $j.pop$
6) $j.pop$ **ena** skip(i)
   $j = \{t \mapsto m + 1 \mid 1 .. m \lhd t = s\}$

Fig. 3. Flow specification for the verification of stack properties

shown on Figure 2 are nodes for the initialisation event start (circle) and implicit deadlock event stop (filled circle). The diagrams like those in Figure 2 (except 5 and 6 which are next-state relations rather than relations over events) are translated into theorems and appear as additional *proof obligations* for the associated Event-B model. A change in the diagram or Event-B model would automatically lead to the recomputation of affected proof obligations. Flow proof obligations are dealt with, like all other proof obligation types, by a combination of automated provers and interactive proof. Like in proofs of model consistency and refinement, the feedback from an undischarged Flow proof obligation may often be interpreted as a suggestion of a diagram change such as an additional assumptions or assertion.

Stack example Let us revisit the stack example from Section 3. The Flow diagram in Figure 3 constructs theorems checking that the algebraic stack properties are satisfied by the Event-B model of stacks. There are five theorems in the diagram. The first three of these check properties $empty(init) = $ TRUE, $empty(push(s, v)) = $ FALSE and $top(push(s, v)) = v$. In the diagram, $init = \varnothing, z \in \mathbb{N}, m \in \mathbb{N}1$ and s is some arbitrary stack state. Property $pop(push(s, v)) = s$ is decomposed into three theorems: $push$ is enabled for an arbitrary state; after $push$, event pop is enabled; pop after $push$ returns the stack into the original state.

It is not difficult to formally demonstrate that the property $(\mathsf{pop} \circ \mathsf{push}) \subseteq \mathrm{id}(\Sigma)$ is implied by the Flow specification. Note that $i.push$ **ena** $j.pop$ $\Leftrightarrow$ $i.push$ **ena** skip(j) $\wedge$ skip(j) **ena** pop. Then the Flow theorems may be trabslated into set theoretic statements as follows.

$$\begin{aligned}
\text{skip}(i) \ \textbf{ena} \ push \ &\Leftrightarrow i \subseteq \mathrm{dom}(push) \\
\text{skip}(j) \ \textbf{ena} \ pop \ &\Leftrightarrow j \subseteq \mathrm{dom}(pop) \\
i.push \ \textbf{ena} \ \text{skip}(j) &\Leftrightarrow push[i] \subseteq j \\
j.pop \ \textbf{ena} \ \text{skip}(i) \ &\Leftrightarrow pop[j] \subseteq i
\end{aligned}$$

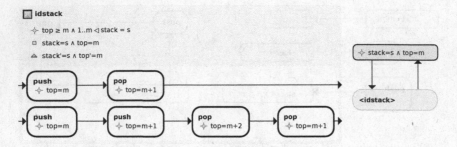

Fig. 4. Folding of idempotent stack scenarious

The statements above is merely the predicate form of a relational join for pop and push : $i \subseteq \mathrm{dom}(push) \wedge push[i] \subseteq j \wedge j \subseteq \mathrm{dom}(pop) \wedge pop[j] \subseteq i \Rightarrow (pop \circ push)[i] \subseteq i$. Since i is an arbitrary stack state, it follows that $(pop \circ push) \subseteq \mathrm{id}(\Sigma)$.

4.3   Structuring

The Flow language offers a number of structuring mechanisms to help in the construction of larger diagrams. Some of the more important of them are addressed by a container diagram element called sub-scenario. A sub-scenario plays differing roles depending on whether it is open or closed.

An *open sub-scenario* defines a property that must be preserved by every event of contained scenarios. Because of the similarity with the way the condition of a safety invariant is formulated, such a property is called an *interval invariant*. Interval invariants are often used when a property is maintained for some part of scenario, and, compared to the propagation of such property via assumptions and assertions, it results in a larger set of (generally) simpler proof obligations as the interval invariant preservation is treated independently from the core theorems of enabling, disabling and feasibility. It also helps to produce more legible diagrams.

A *closed sub-scenario* is a black-box container with a well-defined interface. Externally, it appears to be a simple event (although no such event exists in the associated machine). Internally, it is a Flow specification that may be developed and manipulated independently from the rest of the diagram, subject to some additional constraints. Besides an interval invariant, a closed sub-scenario defines pre- and postconditions. The precondition is a predicate defining a state that is guaranteed to enable at least one of the contained scenarios. A postcondition is a next-state relation characterising the overall effect of a contained scenario as a relation between the initial state of the closed sub-scenario and the state computed by the last event of the scenario. A closed sub-scenario may be linked to a separate diagram file to facilitate collaborative development. An example of a closed sub-scenario is given on diagram in Figure 4 (a green square icon in the left-top distinguishes a closed scenario from an open one). From top to bottom, the predicates are: interval property, precondition and postcondition. In this example, the closed sub-scenario characterises scenarios that, when considered in isolation, have no effect on the stack size and contents. Two trivial examples of such scenarios are defined within the subscenario. On the right-hand size of the diagram is an example of a closed sub-scenario in the role of an event node.

The reason for having the sub-scenario element in two versions rather than two separate element kinds is that the construction of a closed scenario is invariably preceded by a stage when an open sub-scenario is defined. The first step in folding a scenario part into a closed sub-scenario is the identification of a property that holds for the whole part. It is such property that makes a piece of diagram distinct from the rest and, typically, is central to demonstrating the postcondition property.

Sometimes the same constraint or diagram pattern is used repeatedly throughout a scenario. To avoid visual clutter, such repeating parts may be declared separately as *aspects*. An aspect may define a predicate and contain a Flow diagram. The semantics of these are defined by the point of integration. For instance, a link to an aspect from an event node conjuncts the aspect property with existing event constraints.

5   Case Study

As a case study we consider a model of networking file system loosely inspired by the design of NFSv4 [12]. The model focuses on the behaviour of a server accepting requests from clients. Somewhat unusually, requests come in the form of simple programs - operation sequences - executed non-atomically (interleaved with other such requests) on the server side. The server is free to schedule incoming requests as it likes (in a real system this has to do with file locking mechanism). At any moment, there may be any number of running requests but the operations on a file system are always executed atomically. A request may be aborted if the server discovers that the current operation may not be executed. If a request succeeds, the client may receive some data as the request result.

Such system architecture makes it difficult to introduce server functionality as series of small-step refinements and provide strong safety invariants. On the other hand, although it is a relatively small model, there is a number of interesting use cases.

The basic notions of the model are the following.

- m file system (current state)
- Q set of known (accepted) requests
- q set of active (running) requests, $q \subseteq Q$
- $p(t)$ operation vector of request $t \in q$

The state of a running request t is characterised by a pointer $c(t)$ to the current operation in the operation vector $p(t)$, an error flag $e(t)$ and a pair of data register $r_1(t), r_2(t)$ used to store parameter values for the request operations. The first register is normally used to store the path to a file and the second holds data that may be written to file or was read from a file.

register	meaning
$c(t)$	operation counter
$r_1(t)$	data register 1
$r_2(t)$	data register 2
$e(t)$	error flag

The request scheduler of the server is defined by four high-level operations: the acceptance of a new request; starting the execution of a request; request finalisation; request abort. The latter two create and send a reply message to a client that has submitted the request.

operation	semantics	meaning
NEW r, d_1, d_2	$Q' = Q \cup \{r \mapsto d_1 \mapsto d_2\}$	accepts new request (r, d_1, d_2)
FIN r	$q', Q' = \{r\} \lhd q, \{r\} \lhd Q$ ^{2}	finalises the current request
PICK r	$q', c'(t) = q' \cup \{r\}, 1$	prepares a requests for execution
	$p'(r) \mapsto r_1'(r) \mapsto r_2'(r) = Q(r)$	
ABORT r	$q', Q' = \{r\} \lhd q, \{r\} \lhd Q$	aborts the current request

We define a small subset of possible request operations: addition of a new file (not existing already in the system); deletion of an existing file; overwriting the contents of an existing file; file read; file search; and register swap acting as a connector between some other operations. The following is the summary of operations and their beaviour.

name	semantics	condition	meaning
ADD	$m' = m \cup \{r_1 \mapsto r_2\}$	$r_1 \notin m$	adds new file r_1 with contents r_2
DELETE	$m' = \{r_1\} \lhd m$	$r_1 \in m$	deletes existing file r_1
UPDATE	$m' = m \lhd\!\!\!- \{r_1 \mapsto r_2\}$	$r_1 \in m$	overwrites existing file r_1 with r_2
READ	$r_2' = m(r_1)$	$r_1 \in m$	reads an existing file
LOOKUP	$r_2' = r_1$	$r_1 \in m$	looks for a file
XCHG	$r_1', r_2' = r_2, r_1$		register swap

All the operations implicitly update program counter $c(t)$. In addition, when the associated condition is not satisfied, an operation does nothing but raises flag $e(t)$. For the convenience of modelling, operations in $p(t)$ are given in reverse order: $1 \mapsto$ DELETE, $2 \mapsto$ ADD is understood as first ADD and afterwards DELETE.

We have build an Event-B model of the system comprised of four refinement steps. As an illustration, at the last refinement step, the ADD operation is realised by the following event.

add = **any** t **where**
        $q \neq \varnothing\ t \in q$
        $c(t) > 0$
        $p(t)(c(t)) =$ ADD
        $r1(t) \in$ DATA $\setminus$ {NIL}
        $r1(t) \notin \mathrm{dom}(m)$
then
        $m(r1(t)) := r2(t)$
        $c(t) := c(t) - 1 \parallel p(t) := 1 .. c(t) - 1 \lhd p(t)$
end

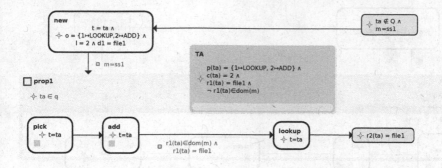

Fig. 5. File look-up after addition

The event picks some request t such that it is not finished ($c(t) > 0$) and its current operation is ADD, i.e., $p(t)(c(t)) = $ ADD. The event also requires that $r1(t)$ contains a valid fresh file name.

In the continuation of the section we discuss three use cases introducing additional verification constraints into the model.

5.1 Use Case 1: File Look-Up After Addition

The use case states that a new file (not previously known in the file system) added with operation ADD may be found with operation LOOKUP immediately after executing ADD. The scenario checks that the notion of file creation implemented by ADD is compatible with the notion of file existence implemented by LOOKUP.

The Flow diagram (see Figure 5) encoding the use case starts with a premise that the file system is in some state $ss1$ and request name ta is not in the pool Q of ready requests. We also rely on the fact that a file with name *file1* does not exists in $m = ss1$. The first step is the creation of a new request made of operations ADD and LOOKUP with register one set to file name *file1*. It is not important what ADD writes in file *file1* thus register two is left unconstrained in event new. To simplify the definition of the remaining of the scenario, the information about the request operations is deposited in an aspect (green box titled TA). Events pick and add integrate this aspect.

After a request is formed it may be selected for execution. This is achieved with event pick. After pick, it is known that ta is some currently running request and this is reflected in the sub-scenario property $ta \in q$. Event add may be executed after pick since, from the aspect definition, it is known that the current operation of the operation vector of ta is pointing at value ADD. The after-state of add defines an updated operation counter $c'(ta)$ pointing at operation LOOKUP. This information is deduced automatically by the provers. Finally, after event lookup we are interested in stating that the second register holds a value corresponding to the name of the created file. To construct the proof it is necessary to propagate the information about the effect of executing ADD in the request. The property we need is that *file1* now exists in the file system and

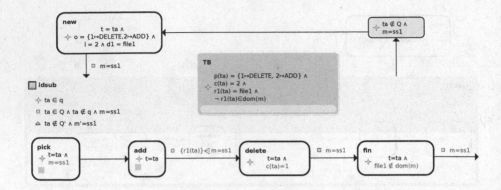

Fig. 6. Idempotent requests template and a sample request specification

the file we are looking for with operation LOOKUP is, in fact, the same file *file1*. Once the property is added the proof goes automatically.

On this and the following Flow diagrams, predicates highlighted in a ligher shade indicate constraints discovered during the proof session. In other words, these elements were not a part of the initial diagram but were necessary to accomplish the proofs.

5.2   Use Case 2: Idempotent Requests

An important family of use cases are those concerned with requests that do not change file system state if executed without an interference from other requests. The only effect of executing such an (idempotent) request is that it disappears from the pool of prepared requests. There is a large number of scenarios that fit this template. The common property is that they start and end with the same file system state ($m = ss1 \wedge m' = ss1$) while some previously inactive request ta would be removed from the pool of ready requests ($ta \in Q \wedge ta \notin Q'$). Since the only way to remove a request from Q is by executing it till completion or abort, these properties (formally, pre- and postconditions) define an execution of an idempotent request. (READ), (LOOKUP) and (XCHG) are the only single-operation requests that should not change the file system state (the former two might result in an aborted request).

As discussed in Section 4.3, a closed sub-scenario defines a family of independent sub-projects that may be, logically and technically, considered in separation from the main Flow diagram. This is an important practical consideration as it allows one to construct large diagrams without overloading the tool infrastructure. There is also a degree of proof economy as the top three arrows are theorems proven just once for the whole family of sub-projects.

One example of an idempotent request is shown on the diagram in Figure 6. It is programmed to delete a file added in the same request. Since the addition operation requires that an added file is not already present in the file, the subsequent deletion of an added file essentially nullifies the effect of addition. With

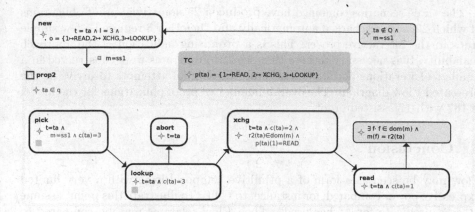

Fig. 7. If a file is found it may be read

condition $\{r1(ta)\} \lhd m = ss1$ we prove that, in the after-state of add, the state of the file system is exactly as before except the appearance of newly added file $r1(ta)$. This property allows us to prove that event delete followed by fin returns the file system into its original state $m = ss1$ by deleting file $r1(ta)$. After finalizing the request (event delete) we prove that the system is back into the state where ta is not a known request and $m = ss1$.

5.3 Use Case 3: File May Be Read If Found

This use case checks the interplay of file existence and file reading. The flow diagram in Figure 7 checks that operation LOOKUP either aborts a request or stores in register $r2$ the name of some existing file and, after copying $r2$ into $r1$, reading of the file with operation READ always succeed and delivers a result in register $r2$ that is the contents of some existing file.

In this scenario, we use branching to select only one possible case of LOOKUP execution. Also, the link between LOOKUP and READ includes an intermediate operation XCHG which requires the propogation via event constraints of the relevant results of lookup execution.

An interesting point is, assuming that operations READ, XCHG and ABORT are correct, that this use case gives a complete characterisation of the behaviour of LOOKUP: the operation either fails or stores in r_2 the name of a file that exists in the file system. This hints at the possibility of synthesis of model parts from detailed Flow diagrams.

The model of the case study (including both the Event-B model and the Flow diagrams) is available to download as a Rodin project [8]. In addition to the core Rodin Platform it is necessary to install the Rodin Flow plugin (instructions may be found in [9]). The diagrams in this paper were produced by printing the diagrams directly from the tool into a PDF file. For presentation purposes, variable names in the diagrams were shortened.

The three scenarios combined have produced 25 non-trivial proof obligations of which 22 were discharged automatically and the other 3 required a few simple steps in the interactive prover. This is a promising indication of the approach scalability. It is necessary to note that scenario diagrams were constructed in a number of iterations and it took several failed proof attempts to arrive at the presented Flow diagrams. The overall number of proof obligations for the model is 167 with 17 assisted proofs.

6   Conclusion

Flow may be seen as a form of a primitive temporal logic with a very limited expressive power compared, for instance, to CTL. To illustrate this point, assume $[f]$ is a predicate for "f has fired". Then Flow relations may be compared to CTL statements as follows:

$$f \text{ ena } g \Rightarrow ([f] \Rightarrow EX[g])$$
$$f \text{ dis } g \Leftrightarrow ([f] \Rightarrow AX \neg[g])$$
$$f \text{ fis } g \Leftarrow ([f] \Rightarrow EX[g])$$

Thus, individual flow theorems characterise the immediate future behaviour and it would take a chain of such theorems to approximate more interesting statements. For instance, from a fact that f enables g and g enables h one may conclude that after f there is a path leading to h: $(f \text{ ena } g \wedge g \text{ ena } h) \Rightarrow ([f] \Rightarrow EF[h])$. In principle, a statement like $[f] \Rightarrow EF[g]$ is expressible as a flow scenario for any two events f and g[3]. On the other hand, in most cases, it should be impractical to approximate $[f] \Rightarrow AF[g]$ and $[f] \Rightarrow AG[g]$ as this requires formulation of possible continuation scenarios and a proof (possibly, via **dis**) that no other continuation exists.

The most closely related work is a study of liveness-style theorems for the Classical B [2]. The work introduces a number of notation extensions to construct proofs about 'dynamic' properties of models - properties that span over several event executions. Like in Flow, the formulation of reachability property requires spelling out a path that would lead to its satisfaction. One advantage of our approach is in the use of graphs to construct complex theorems from simple ones and the propagation of properties along the graph structure. The latter results in interactive modelling/proof sessions where proof feedback leads to small, incremental changes in the diagram.

There are a number of approaches [11,6,4,13] on combining process algebraic specification with event-based formalisms such as Event-B and Action Systems. The fundamental difference is that Flow does not introduce behavioural constraints and is simply a high-level notation for writing certain kind of theorems. It would be interesting to explore how explicit control flow information present in a process algebraic model part may affect the applicability and the practice of the Flow approach.

---

[3] This requires a side condition that any loops are proven convergent in the Event-B part.

Acknowledgments. This work is supported by FP7 ICT DEPLOY Project and the EPSRC/UK TrAmS platform grant.

References

1. Abrial, J.-R.: Modelling in Event-B. Cambridge University Press, Cambridge (2010)
2. Abrial, J.-R., Mussat, L.: Introducing Dynamic Constraints in B. In: Bert, D. (ed.) B 1998. LNCS, vol. 1393, pp. 83–128. Springer, Heidelberg (1998)
3. Event-B.org. Event-B model repository (2011), http://deploy-eprints.ecs.soton.ac.uk/view/type/rodin=5Farchive.html
4. Fischer, C., Wehrheim, H.: Model-Checking CSP-OZ Specifications with FDR. In: Araki, A., Galloway, A., Taguchi, K. (eds.) IFM 1999: Proceedings of the 1st International Conference on Integrated Formal Methods, London, UK, pp. 315–334. Springer, Heidelberg (1999)
5. Hurlbut, R.R.: A survey of approaches for describing and formalizing use cases. Technical report, Expertech, Ltd. (1997)
6. Butler, M., Leuschel, M.: Combining CSP and B for Specification and Property Verification, pp. 221–236 (2005)
7. Mendoza-Grado, V.M.: Formal Verification of Use Cases. In: Requirements Engineering: Use Cases and More (1995)
8. Flow Models of stack and NFS. Event B/Flow specification (2011), http://iliasov.org/usecase/nfs.zip
9. Plugin, F.: Event-B wiki page (2011), http://wiki.event-b.org/index.php/Flows
10. The RODIN platform, http://rodin-b-sharp.sourceforge.net/
11. Treharne, H., Schneider, S., Bramble, M.: Composing Specifications Using Communication. In: Bert, D., Bowen, J.P., King, S. (eds.) ZB 2003. LNCS, vol. 2651, pp. 58–78. Springer, Heidelberg (2003)
12. NFSv4 web page. Network File System Version 4 (2011), http://datatracker.ietf.org/wg/nfsv4/
13. Woodcock, J., Cavalcanti, A.: The Semantics of Circus. In: Bert, D., Bowen, J.P., Henson, M.C., Robinson, K. (eds.) B 2002 and ZB 2002. LNCS, vol. 2272, pp. 184–203. Springer, Heidelberg (2002)

Quantitative Verification of System Safety in Event-B

Anton Tarasyuk[1,2], Elena Troubitsyna[2], and Linas Laibinis[2]

[1] Turku Centre for Computer Science
[2] Åbo Akademi University
Joukahaisenkatu 3-5, 20520 Turku, Finland
{anton.tarasyuk,elena.troubitsyna,linas.laibinis}@abo.fi

Abstract. Certification of safety-critical systems requires formal verification of system properties and behaviour as well as quantitative demonstration of safety. Usually, formal modelling frameworks do not include quantitative assessment of safety. This has a negative impact on productivity and predictability of system development. In this paper we present an approach to integrating quantitative safety analysis into formal system modelling and verification in Event-B. The proposed approach is based on an extension of Event-B, which allows us to perform quantitative assessment of safety within proof-based verification of system behaviour. This enables development of systems that are not only correct but also safe by construction. The approach is demonstrated by a case study – an automatic railway crossing system.

1 Introduction

Safety is a property of a system to not endanger human life or environment [4]. To guarantee safety, designers employ various rigorous techniques for formal modelling and verification. Such techniques facilitate formal reasoning about system correctness. In particular, they allow us to guarantee that a safety invariant – a logical representation of safety – is always preserved during system execution. However, real safety-critical systems, i.e., the systems whose components are susceptible to various kinds of faults, are not "absolutely" safe. In other words, certain combinations of failures may lead to an occurrence of a hazard – a potentially dangerous situation breaching safety requirements. While designing and certifying safety-critical systems, we should demonstrate that the probability of a hazard occurrence is acceptably low. In this paper we propose an approach to combining formal system modelling and quantitative safety analysis.

Our approach is based on a probabilistic extension of Event-B [22]. Event-B is a formal modelling framework for developing systems correct-by-construction [3,1]. It is actively used in the EU project Deploy [6] for modelling and verifying of complex systems from various domains including railways. The Rodin platform [20] provides the designers with an automated tool support that facilitates formal verification and makes Event-B relevant in an industrial setting.

E.A. Troubitsyna (Ed.): SERENE 2011, LNCS 6968, pp. 24–39, 2011.

The main development technique of Event-B is refinement – a top-down process of gradual unfolding of the system structure and elaborating on its functionality. In this paper we propose design strategies that allow the developers to structure safety requirements according to the system abstraction layers. Essentially, such an approach can be seen as a process of extracting a fault tree – a logical representation of a hazardous situation in terms of the primitives used at different abstraction layers. Eventually, we arrive at the representation of a hazard in terms of failures of basic system components. Since our model explicitly contains probabilities of component failures, standard calculations allow us to obtain a probabilistic evaluation of a hazard occurrence. As a result, we obtain an algebraic representation of the probability of safety violation. This probability is defined using the probabilities of system component failures. To illustrate our approach, we present a formal development and safety analysis of a radio-based railway crossing. We believe the proposed approach can potentially facilitate development, verification and assessment of safety-critical systems.

The rest of the paper is organised as follows. In Section 2 we describe our formal modelling framework – Event-B, and briefly introduce its probabilistic extension. In Section 3 we discuss a general design strategy for specifying Event-B models amenable for probabilistic analysis of system safety. In Section 4 we demonstrate the presented approach by a case study. Finally, Section 5 presents an overview of the related work and some concluding remarks.

2   Modelling in Event-B

The B Method [2] is an approach for the industrial development of highly dependable software. The method has been successfully used in the development of several complex real-life applications [19,5]. Event-B is a formal framework derived from the B Method to model parallel, distributed and reactive systems. The Rodin platform provides automated tool support for modelling and verification in Event-B. Currently Event-B is used in the EU project Deploy to model several industrial systems from automotive, railway, space and business domains.

Event-B Language and Semantics. In Event-B, a system model is defined using the notion of an *abstract state machine* [18]. An abstract state machine encapsulates the model state, represented as a collection of model variables, and defines operations on this state. Therefore, it describes the dynamic part of the modelled system. A machine may also have an accompanying component, called *context*, which contains the static part of the system. In particular, It can include user-defined carrier sets, constants and their properties given as a list of model axioms. A general form of Event-B models is given in Fig. 1.

The machine is uniquely identified by its name M. The state variables, v, are declared in the **Variables** clause and initialised in the *Init* event. The variables are strongly typed by the constraining predicates I given in the **Invariants** clause. The invariant clause also contains other predicates defining properties that must be preserved during system execution.

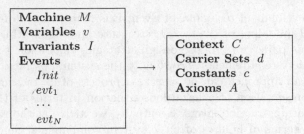

Fig. 1. Event-B machine and context

Action (S)	$BA(S)$
$x := E(x, y)$	$x' = E(x, y) \ \wedge \ y' = y$
$x :\in Set$	$\exists z \cdot (z \in Set \wedge x' = z) \ \wedge \ y' = y$
$x :\mid Q(x, y, x')$	$\exists z \cdot (Q(x, z, y) \wedge x' = z) \ \wedge \ y' = y$

Fig. 2. Before-after predicates

The dynamic behaviour of the system is defined by the set of atomic events specified in the **Events** clause. Generally, an event can be defined as follows:

$$evt \ \widehat{=} \ \textbf{any} \ a \ \textbf{where} \ g \ \textbf{then} \ S \ \textbf{end},$$

where a is the list of local variables, the guard g is a conjunction of predicates over the local variables a and state variables v, while the action S is a state assignment. If the list a is empty, an event can be described simply as

$$evt \ \widehat{=} \ \textbf{when} \ g \ \textbf{then} \ S \ \textbf{end}.$$

The occurrence of events represents the observable behaviour of the system. The guard defines the conditions under which the action can be executed, i.e., when the event is *enabled*. If several events are enabled at the same time, any of them can be chosen for execution nondeterministically. If none of the events is enabled then the system deadlocks.

In general, the action of an event is a parallel composition of variable assignments. The assignments can be either deterministic or non-deterministic. A deterministic assignment, $x := E(x, y)$, has the standard syntax and meaning. A nondeterministic assignment is denoted either as $x :\in Set$, where Set is a set of values, or $x :\mid Q(x, y, x')$, where Q is a predicate relating initial values of x, y to some final value of x'. As a result of such a non-deterministic assignment, x can get any value belonging to Set or satisfying Q.

The semantics of Event-B actions is defined using so-called before-after (BA) predicates [3,18]. A BA predicate describes a relationship between the system states before and after execution of an event, as shown in Fig. 2. Here x and y are disjoint lists of state variables, and x', y' represent their values in the after-state. The semantics of a whole Event-B model is formulated as a number of *proof obligations*, expressed in the form of logical sequents. The full list of proof obligations can be found in [3].

Probabilistic Event-B. In our previous work [22] we have have extended the Event-B modelling language with a new operator – *quantitative probabilistic choice*, denoted $\oplus|$. It has the following syntax

$$x \oplus| \ x_1 \ @ \ p_1; \ldots; x_n \ @ \ p_n,$$

where $\sum_{i=1}^{n} p_i = 1$. It assigns to the variable x a new value x_i with the corresponding non-zero probability p_i. The quantitative probabilistic choice (assignment) allows us to precisely represent the probabilistic information about how likely a particular choice is made. In other words, it behaves according to some known probabilistic distribution.

We have restricted the use of the new probabilistic choice operator by introducing it only to replace the existing demonic one. This approach has also been adopted by Hallerstede and Hoang, who have proposed extending the Event-B framework with *qualitative probabilistic choice* [10]. It has been shown that any probabilistic choice statement always refines its demonic nondeterministic counterpart [13]. Hence, such an extension is not interfering with the established refinement process. Therefore, we can rely on the Event-B proof obligations to guarantee functional correctness of a refinement step. Moreover, the probabilistic information introduced in new quantitative probabilistic choices can be used to stochastically evaluate certain non-functional system properties.

For instance, in [22] we have shown how the notion of Event-B refinement can be strengthened to quantitatively demonstrate that the refined system is more reliable than its abstract counterpart. In this paper we aim at enabling quantitative safety analysis within Event-B development.

3   Safety Analysis in Event-B

In this paper we focus on modelling of highly dynamic reactive control systems. Such systems provide instant control actions as a result of receiving stimuli from the controlled environment. Such a restriction prevents the system from executing automated error recovery, i.e. once a component fails, its failure is considered to be permanent and the system ceases its automatic functioning.

Generally, control systems are built in a layered fashion and reasoning about their behaviour is conducted by unfolding layers of abstraction. Deductive safety analysis is performed in a similar way. We start by identifying a hazard – a dangerous situation associated with the system. By unfolding the layers of abstraction we formulate the hazard in terms of component states of different layers.

In an Event-B model, a hazard can be naturally defined as a predicate over the system variables. Sometimes, it is more convenient to reformulate a hazard as a dual safety requirement (property) that specifies a proper behaviour of a system in a hazardous situation. The general form of such a safety property is:

$$SAF \ \hat{=} \ H(v) \Rightarrow K(v),$$

where the predicate $H(v)$ specifies a hazardous situation and the predicate $K(v)$ defines the safety requirements in the terms of the system variables and states.

The essential properties of an Event-B model are usually formulated as invariants. However, to represent system behaviour realistically, our specification should include modelling of not only normal behaviour but also component failure occurrence. Since certain combinations of failures will lead to hazardous situations, we cannot guarantee "absolute" preservation of safety invariants. Indeed, the goal of development of safety-critical systems is to guarantee that the probability of violation of safety requirements is sufficiently small.

To assess the preservation of a desired safety property, we will unfold it (in the refinement process) until it refers only to concrete system components that have direct impact on the system safety. To quantitatively evaluate this impact, we require that these components are probabilistically modelled in Event-B using the available information about their reliability. Next we demonstrate how the process of unfolding the safety property from the abstract to the required concrete representation can be integrated into our formal system development.

Often, functioning of a system can be structured according to a number of *execution stages*. There is a specific component functionality associated with each stage. Since there is no possibility to replace or repair failed system components, we can divide the process of quantitative safety assessment into several consecutive steps, where each step corresponds to a particular stage of the system functioning. Moreover, a relationship between different failures of components and the system behaviour at a certain execution stage is preserved during all the subsequent stages. On the other hand, different subsystems can communicate with each other, which leads to possible additional dependencies between system failures (not necessarily within the same execution stage). This fact significantly complicates quantitative evaluation of the system safety.

We can unfold system safety properties either in a *backward* or in a *forward* way. In the backward unfolding we start from the last execution stage preceding the stage associated with the potentially hazardous situation. In the forward one we start from the first execution stage of the system and continue until the last stage just before the hazardous situation occurs. In this paper we follow the former approach. The main idea is to perform a stepwise analysis of any possible behaviour of all the subsystems at every execution stage preceding the hazardous situation, while gradually unfolding the abstract safety property in terms of new (concrete) variables representing faulty components of the system.

Specifically, in each refinement step, we have to establish the relationship between the newly introduced variables and the abstract variables present in the safety property. A standard way to achieve this is to formulate the required relationship as a number of safety invariants in Event-B. According to our development strategy, each such invariant establishes a connection between abstract and more concrete variables that have an impact on system safety. Moreover, the preservation of a safety invariant is usually verified for a particular subsystem at a specific stage. Therefore, we can define a general form of such an invariant:

$$I_s(v, u) \mathrel{\widehat{=}} F(v) \Rightarrow (K(v) \Leftrightarrow L(u)),$$

where the predicate F restricts the execution stage and the subsystems involved, while the predicate $K \Leftrightarrow L$ relates the values of the newly introduced variables

u with the values the abstract variables v present in the initially defined safety property or/and in the safety invariants defined in the previous refinement steps.

To calculate the probability of preservation of the safety property, the refinement process should be continued until all the abstract variables, used in the definition of the system safety property, are related to the concrete, probabilistically updated variables, representing various system failures or malfunctioning. The process of probability evaluation is rather straightforward and based on basic definitions and rules for calculating probabilities (see [7] for instance).

Let us consider a small yet generic example illustrating the calculation of probability using Event-B safety invariants. We assume that the safety property SAF is defined as above. In addition, let us define two safety invariants – I_s and J_s – introduced in two subsequent refinement steps. More specifically,

$$I_s \cong F \Rightarrow (K(v) \Leftrightarrow L_1(u_1) \vee L_2(u_2)) \text{ and } J_s \cong \tilde{F} \Rightarrow (L_2(u_2) \Leftrightarrow N(w)),$$

where $u_1 \subset u, u_1 \neq \emptyset$ are updated probabilistically in the first refinement, while $u_2 = u \setminus u_1$ are still abstract in the first refinement machine and related by J_s to the probabilistically updated variables w in the following one. Let us note that the predicate $\tilde{F}$ must define the earlier stage of the system than the predicate F does. Then the probability that the safety property SAF is preserved is

$$P_{SAF} = P\{K(v)\} = P\{L_1(u_1) \vee L_2(u_2)\} = P\{L_1(u_1) \vee N(w)\} =$$
$$P\{L_1(u_1)\} + P\{N(w)\} - P\{L_1(u_1) \wedge N(w)\},$$
where
$$P\{L_1(u_1) \wedge N(w)\} = P\{L_1(u_1)\} \cdot P\{N(w)\}$$
in the case of independent L_1 and N, and
$$P\{L_1(u_1) \wedge N(w)\} = P\{L_1(u_1)\} \cdot P\{N(w) \mid L_1(u_1)\}$$
otherwise. Note that the predicate $H(v)$ is not participating in the calculation of P_{SAF} directly. Instead, it defines "the time and the place" when and where the values of the variables u and v should be considered, and, as long as it specifies the hazardous situation following the stages defined by F and $\tilde{F}$, it can be understood as the *post-state* for all the probabilistic events.

In the next section we will demonstrate the approach presented above by a case study – an automatic railway crossing system.

4   Case Study

To illustrate safety analysis in the probabilistically enriched Event-B method, in this section we present a quantitative safety analysis of a radio-based railway crossing. This case study is included into priority program 1064 of the German Research Council (DFG) prepared in cooperation with Deutsche Bahn AG. The main difference between the proposed technology and traditional control systems of railway crossings is that signals and sensors on the route are replaced by radio communication and software computations performed at the train and railway crossings. Formal system modelling of such a system has been undertaken previously [16,15]. However, the presented methodology is focused on logical (qualitative) reasoning about safety and does not include quantitative safety analysis. Below we demonstrate how to integrate formal modelling and probabilistic safety analysis.

Let us now briefly describe the functioning of a radio-based railway crossing system. The train on the route continuously computes its position. When it approaches a crossing, it broadcasts a *close* request to the crossing. When the railway crossing receives the command *close*, it performs some routine control to ensure safe train passage. It includes switching on the traffic lights, that is followed by an attempt to close the barriers. Shortly before the train reaches the *latest braking point*, i.e., the latest point where it is still possible for the train to stop safely, it requests the *status* of the railway crossing. If the crossing is secured, it responds with a *release* signal, which indicates that the train may pass the crossing. Otherwise, the train has to brake and stop before the crossing. More detailed requirements can be found in [16] for instance.

In our development we abstract away from modelling train movement, calculating train positions and routine control by the railway crossing. Let us note that, any time when the train approaches to the railway crossing, it sequentially performs a number of predefined operations:

- it sends the *close* request to the crossing controller;
- after a delay it sends the *status* request;
- it awaits for an answer from the crossing controller.

The crossing controller, upon receiving the close request, tries to close the barriers and, if successful, sends the *release* signal to the train. Otherwise, it does not send any signal and in this case the train activates the emergency brakes. Our safety analysis focused on defining the hazardous events that may happen in such a railway crossing system due to different hardware and/or communication failures, and assess the probability of the hazard occurrences. We make the following fault assumptions:

- the radio communication is unreliable and can cause messages to be lost;
- the crossing barrier motors may fail to start;
- the positioning sensors that are used by the crossing controller to determine a physical position of the barriers are unreliable;
- the train emergency brakes may fail.

The abstract model. We start our development with identification of all the high-level subsystems we have to model. Essentially, our system consists of two main components – the train and the crossing controller. The system environment is represented by the physical position of the train. Therefore, each control cycle consists of three main phases – *Env*, *Train* and *Crossing*. To indicate the current *phase*, the eponymous variable is used.

The type modelling abstract train positions is defined as the enumerated set of nonnegative integers $POS\_SET = \{0, CRP, SRP, SRS, DS\}$, where $0 < CRP < SRP < SRS < DS$. Each value of POS_SET represents a specific position of the train. Here 0 stands for some initial train position outside the communication area, CRP and SRP stand for the close and status request points, and SRS and DS represent the safe reaction and danger spots respectively. The actual train position is modelled by the variable $train\_pos \in POS\_SET$. In

```
Machine RailwayCrossing
Variables  train_pos, phase, emrg_brakes, bar₁, bar₂
Invariants  · · ·
Events   · · ·
  UpdatePosition₁ ≙
    when phase = Env ∧ train_pos < DS ∧ emrg_brakes = FALSE
    then
        train_pos := min({p | p ∈ POS_SET ∧ p > train_pos}) || phase := Train
    end
  UpdatePosition₂ ≙
    when phase = Env ∧
          ((train_pos = DS ∧ emrg_brakes = FALSE) ∨ emrg_brakes = TRUE)
    then
      skip
    end
  TrainIdle ≙
    when phase = Train ∧ train_pos ≠ SRS
    then
      phase := Crossing
    end
  TrainReact ≙
    when phase = Train ∧ train_pos = SRS
    then
        emrg_brakes :∈ BOOL || phase := Crossing
    end
  CrossingBars ≙
    when phase = Crossing ∧ train_pos = CRP
    then
      bar₁, bar₂ :∈ BAR_POS || phase := Env
    end
  CrossingIdle ≙
    when phase = Crossing ∧ train_pos ≠ CRP
    then
      phase := Env
    end
```

Fig. 3. Railway crossing: the abstract machine

addition, we use the boolean variable *emrg_brakes* to model the status of the train emergency brakes. We assume that initially they are not triggered, i.e., $emrg\_brakes = FALSE$.

The crossing has two barriers – one at each side of the crossing. The status of the barriers is modelled by the variables bar_1 and bar_2 that can take values *Opened* and *Closed*. We assume that both barriers are initially open.

The initial abstract machine *RailwayCrossing* is presented in Fig. 3. We omit showing here the *Initialisation* event and the **Invariants** clause (it merely defines the types of variables). Due to lack of space, in the rest of the section we will also present only some selected excerpts of the model. The full Event-B specifications of the *Railway crossing system* can be found in [21].

In *RailwayCrossing* we consider only the basic functionality of the system. Two events *UpdatePosition₁* and *UpdatePosition₂* are used to abstractly model train movement. The first event models the train movement outside the danger spot by updating the train abstract position according to the next value of the *POS_SET*. *UpdatePosition₂* models the train behaviour after it has passed the last braking point or when it has stopped in the safe reaction spot. Essentially, this event represents the system termination (both safe and unsafe cases), which is modelled as infinite stuttering. Such an approach for modelling of the train movement is sufficient since we only analyse system behaviour within the train-crossing communication area, i.e., the area that consists of the close and status request points, and the safe reaction spot. A more realistic approach for modelling of the train movement is out of the scope of our safety analysis.

For the crossing controller, we abstractly model closing of the barriers by the event *CrossingBar*, which non-deterministically assigns the variables bar_1 and bar_2 from the set BAR_POS. Let us note that in the abstract machine the crossing controller immediately knows when the train enters the close request area and makes an attempt to close the barriers. In further refinement steps we eliminate this unrealistic abstraction by introducing communication between the train and the crossing controller. In addition, in the *Train* phase the event *TrainReact* models triggering of the train brakes in the safe reaction spot.

The hazard present in the system is the situation when the train passes the crossing while at least one barrier is not closed. In terms of the introduced system variables and their states it can defined as follows:

$$train\_pos = DS \wedge (bar_1 = Opened \vee bar_2 = Opened).$$

In a more traditional (for Event-B invariants) form, this hazard can be dually reformulated as the following safety property:

$$train\_pos = SRS \wedge phase = Crossing \Rightarrow$$
$$(bar_1 = Closed \wedge bar_2 = Closed) \vee emrg\_brakes = TRUE. \quad (1)$$

This safety requirement can be interpreted as follows: after the train, being in the safe reaction spot, reacts on signals from the crossing controller, the system is in the safe state only when both barriers are closed or the emergency brakes are activated. Obviously, this property cannot be formulated as an Event-B invariant – it might be violated due to possible communication and/or hardware failures. Our goal is to assess the probability of violation of the safety property (1). To achieve this, during the refinement process, we have to unfold (1) by introducing representation of all the system components that have impact on safety. Moreover, we should establish a relationship between the variables representing these components and the abstract variables present in (1).

The first refinement. In the first refinement step we examine in detail the system behaviour at the safe reaction spot – the last train position preceding the danger spot where the hazard may occur. As a result, the abstract event *TrainReact* is refined by three events *TrainRelease$_1$*, *TrainRelease$_2$* and *TrainStop* that represent reaction of the train on the presence or absence of the release signal from the crossing controller. The first two events are used to model the situations when the release signal has been successfully delivered or lost respectively. The last one models the situation when the release signal has not been sent due to some problems at the crossing controller side. Please note that since the events *TrainRelease$_2$* and *TrainStop* perform the same actions, i.e., trigger the emergency brakes, they differ only in their guards.

The event *CrossingStatusReq* that "decides" whether to send or not to send the release signal is very abstract at this stage – it does not have any specific guards except those that define the system phase and train position. Moreover, the variable *release_snd* is updated in the event body non-deterministically. To

```
Machine RailwayCrossing_R1
Variables ..., release_snd, release_rcv, emrg_brakes_failure, release_com_failure, ...
Invariants  ···
Events  ···
  TrainRelease₁ ≙
    when phase = Train ∧ train_pos = SRS ∧ release_snd = TRUE
         release_comm_failure = FALSE ∧ deceleration = FALSE ∧ comm_ct = FALSE
    then
         emrg_brakes := FALSE || release_rcv := TRUE || phase := Crossing
    end
  TrainRelease₂ ≙
    when phase = Train ∧ train_pos = SRS ∧ release_snd = TRUE
         release_comm_failure = TRUE ∧ deceleration = FALSE ∧ comm_ct = FALSE
    then
         emrg_brakes :| emrg_brakes' ∈ BOOL ∧ (emrg_brakes' = TRUE ⟺
                                                      emrg_brakes_failure = FALSE)
         release_rcv := TRUE || phase := Crossing
    end
  TrainStop ≙
    when phase = Train ∧ train_pos = SRS ∧ release_snd = FALSE ∧ deceleration = FALSE
    then
         ...
    end
  CrossingStatusReq ≙
    when phase = Crossing ∧ train_pos = SRP
    then
         release_snd :∈ BOOL || phase := Env
    end
  ReleaseComm ≙
    when phase = Train ∧ train_pos = SRS ∧ release_snd = TRUE ∧ comm_ct = TRUE
    then
         release_comm_failure ⊕| TRUE @ p₁; FALSE @ 1−p₁ || comm_ct := FALSE
    end
  TrainDec ≙
    when phase = Train ∧ train_pos = SRS ∧ deceleration = TRUE
    then
         emrg_brakes_failure ⊕| TRUE @ p₄; FALSE @ 1−p₄ || deceleration := FALSE
    end
```

Fig. 4. Railway crossing: first refinement

model the failures of communication and emergency brakes, we introduce two new events with *probabilistic* bodies – the events *ReleaseComm* and *TrainDec* correspondingly. For convenience, we consider communication as a part of the receiving side behaviour. Thus the release communication failure occurrence is modelled in the *Train* phase while the train being in the *SRS* position. Some key details of the Event-B machine *RailwayCrossing_R1* that refines the abstract machine *RailwayCrossing* are shown in Fig. 4.

The presence of concrete variables representing unreliable system components in *RailwayCrossing_R1* allows us to formulate two safety invariants (saf_inv_1 and saf_inv_2) that glue the abstract variable $emrg\_brakes$ participating in the safety requirement (1) with the (more) concrete variables $release\_rcv$, $emrg\_brakes\_failure$, $release\_snd$ and $release\_com\_failure$.

$$\mathbf{saf\_inv_1} : train\_pos = SRS \wedge phase = Crossing \Rightarrow (emrg\_brakes = TRUE \Leftrightarrow$$
$$release\_rcv = FALSE \wedge emrg\_brakes\_failure = FALSE)$$

$$\mathbf{saf\_inv_2} : train\_pos = SRS \wedge phase = Crossing \Rightarrow (release\_rcv = FALSE \Leftrightarrow$$
$$release\_snd = FALSE \vee release\_comm\_failure = TRUE)$$

We split the relationship between the variables into two invariant properties just to improve the readability and make the invariants easier to understand. Obviously, since the antecedents of both invariants coincide, one can easily merge them together by replacing the variable $release\_rcv$ in saf_inv_1 with the right hand side of the equivalence in the consequent of saf_inv_1. Please note that the variable $release\_snd$ corresponds to a certain combination of system actions and hence should be further unfolded during the refinement process.

The second refinement. In the second refinement step we further elaborate on the system functionality. In particular, we model the request messages that the train sends to the crossing controller, as well as sensors that read the position of the barriers. Selected excerpts from the second refinement machine $RailwayCrossing\_R2$ are shown in Fig. 5. To model sending of the close and status requests by the train, we refine the event $TrainIdle$ by two simple events $TrainCloseReq$ and $TrainStatusReq$ that activate sending of the close and status requests at the corresponding stages. In the crossing controller part, we refine the event $CrossingBars$ by the event $CrossingCloseReq$ that sets the actuators closing the barriers in response to the close request from the train. Clearly, in the case of communication failure occurrence during the close request transmission, both barriers remain open.

Moreover, the abstract event $CrossingStatusReq$ is refined by two events $CrossingStatusReq_1$ and $CrossingStatusReq_2$ to model a reaction of the crossing controller on the status request. The former event is used to model the situation when the close request has been successfully received (at the previous stage) and the latter one models the opposite situation. Notice that in the refined event $CrossingStatusReq_1$ the controller sends the release signal only when it has received both request signals and identified that both barriers are closed. This interconnection is reflected in the safety invariant saf_inv_3.

$$\textbf{saf\_inv}_3 : train\_pos = SRP \land phase = Env \Rightarrow$$
$$(release\_snd = TRUE \Leftrightarrow close\_req\_rcv = TRUE \land$$
$$status\_req\_rcv = TRUE \land sensor_1 = Closed \land sensor_2 = Closed)$$

Here the variables $sensor_1$ and $sensor_2$ represent values of the barrier positioning sensors. Let us remind that the sensors are unreliable and can return the actual position of the barriers incorrectly. Specifically, the sensors can get stuck at their previous values or spontaneously change the values to the opposite ones. In addition, to model the communication failures, we add two new events $CloseComm$ and $StatusComm$. These events are similar to the $ReleaseComm$ event of the $RailwayCrossing\_R1$ machine. Rather intuitive dependencies between the train requests delivery and communication failure occurrences are defined by a pair of safety invariants saf_inv_4 and saf_inv_5.

$$\textbf{saf\_inv}_4 : train\_pos = SRP \land phase = Env \Rightarrow$$
$$(status\_req\_rcv = TRUE \Leftrightarrow status\_com\_failure = FALSE)$$

$$\textbf{saf\_inv}_5 : train\_pos = CRP \land phase = Env \Rightarrow$$
$$(close\_req\_rcv = TRUE \Leftrightarrow close\_com\_failure = FALSE)$$

Machine *RailwayCrossing_R2*
Variables $\ldots, close\_snd, close\_rcv, status\_snd, status\_rcv,$
$\qquad\qquad\qquad close\_com\_failure, status\_com\_failure, sensor_1, sensor_2 \ldots$
Invariants $\cdots$
Events $\cdots$

  $TrainCloseReq \; \widehat{=}$
  when $phase = Train \wedge train\_pos = CRP$
  then
    $close\_req\_snd := TRUE \,\|\, phase := Crossing$
  end
    $\cdots$

  $CrossingCloseReq \; \widehat{=}$
  when $phase = Crossing \wedge close\_req\_snd = TRUE \wedge comm\_tc = FALSE$
  then
    $bar_1, bar_2 :\! | \; bar_1' \in BAR\_POS \wedge bar_2' \in BAR\_POS \wedge$
    $\qquad\qquad\qquad (close\_comm\_failure = TRUE \Rightarrow bar_1' = Opened \wedge bar_2' = Opened)$
    $close\_req\_rcv :\! | \; close\_req\_rcv' \in BOOL \wedge$
    $\qquad\qquad\qquad (close\_req\_rcv' = TRUE \Leftrightarrow close\_comm\_failure = FALSE)$
    $comm\_tc := TRUE \,\|\, phase := Env$
  end

  $CrossingStatusReq_1 \; \widehat{=}$
  when $phase = Crossing \wedge status\_req\_snd = TRUE \wedge close\_req\_rcv = TRUE \wedge$
    $\qquad sens\_reading = FALSE \wedge comm\_tc = FALSE$
  then
    $release\_snd :\! | \; release\_snd' \in BOOL \wedge (release\_snd' = TRUE \Leftrightarrow$
    $\qquad\qquad status\_comm\_failure = FALSE \wedge sensor_1 = Closed \wedge sensor_2 = Closed)$
    $status\_req\_rcv :\! | \; status\_req\_rcv' \in BOOL \wedge$
    $\qquad\qquad\qquad (status\_req\_rcv' = TRUE \Leftrightarrow status\_comm\_failure = FALSE)$
    $comm\_tc := TRUE \,\|\, phase := Env$
  end
    $\cdots$

  $ReadSensors \; \widehat{=}$
  when $phase = Crossing \wedge status\_req\_snd = TRUE \wedge sens\_reading = TRUE$
  then
    $sensor_1 :\in \{bar_1, bnot(bar_1)\} \,\|\, sensor_2 :\in \{bar_2, bnot(bar_2)\} \,\|\, sens\_reading := FALSE$
  end

Fig. 5. Railway crossing: second refinement

The third refinement. In the third Event-B machine *RailwayCrossing_R3*, we refine the remaining abstract representation of components mentioned in the safety requirement (1), i.e., modelling of the barrier motors and positioning sensors. We introduce the new variables $bar\_failure_1$, $bar\_failure_2$, $sensor\_failure_1$ and $sensor\_failure_2$ to model the hardware failures. These variables are assigned probabilistically in the newly introduced events *BarStatus* and *SensorStatus* in the same way as it was done for the communication and emergency brakes failures in the first refinement. We refine *CrossingCloseReq* and *ReadSensors* events accordingly. Finally, we formulate four safety invariants saf_inv_6, saf_inv_7, saf_inv_8 and saf_inv_9 to specify the correlation between the physical position of the barriers, the sensor readings, and the hardware failures.

saf_inv$_6$: $train\_pos = CRP \wedge phase = Env \Rightarrow (bar_1 = Closed \Leftrightarrow$
$\qquad\qquad\qquad bar\_failure_1 = FALSE \wedge close\_comm\_failure = FALSE)$

saf_inv$_7$: $train\_pos = CRP \wedge phase = Env \Rightarrow (bar_2 = Closed \Leftrightarrow$
$\qquad\qquad\qquad bar\_failure_2 = FALSE \wedge close\_comm\_failure = FALSE)$

saf_inv$_8$: $train\_pos = SRP \land phase = Env \Rightarrow (sensor_1 = Closed \Leftrightarrow$

$$((bar_1 = Closed \land sensor\_failure_1 = FALSE) \lor$$

$$(bar_1 = Opened \land sensor\_failure_1 = TRUE)))$$

saf_inv$_9$: $train\_pos = SRP \land phase = Env \Rightarrow (sensor_2 = Closed \Leftrightarrow$

$$((bar_2 = Closed \land sensor\_failure_2 = FALSE) \lor$$

$$(bar_2 = Opened \land sensor\_failure_2 = TRUE)))$$

The first two invariants state that the crossing barrier can be closed (in the post-state) only when the controller has received the close request and the barrier motor has not failed to start. The second pair of invariants postulates that the positioning sensor may return the value *Closed* in two cases – when the barrier is closed and the sensor works properly, or when the barrier has got stuck while opened and the sensor misreads its position.

Once we have formulated the last four safety invariants, there is no longer any variable, in the safety property (1), that cannot be expressed via some probabilistically updated variables introduced during the refinement process. This allows us to calculate the probability P_{SAF} that (1) is preserved:

$$P_{SAF} = P\{(bar_1 = Closed \land bar_2 = Closed) \lor emrg\_brakes = TRUE\} =$$

$$P\{bar_1 = Closed \land bar_2 = Closed\} + P\{emrg\_brakes = TRUE\} -$$

$$P\{bar_1 = Closed \land bar_2 = Closed\} \cdot$$

$$P\{emrg\_brakes = TRUE \mid bar_1 = Closed \land bar_2 = Closed\}.$$

Let us recall that we have idenified four different types of failures in our system – the communication failure, the failure of the barrier motor, the sensor failure and emergency brakes failure. We suppose that the probabilities of all these failures are constant and equal to p_1, p_2, p_3 and p_4 correspondingly. The first probability presented in the sum above can be trivially calculated based on the safety invariants saf_inv_7 and saf_inv_8:

$$P\{bar_1 = Closed \land bar_2 = Closed\} =$$

$$P\{bar\_failure_1 = FALSE \land bar\_failure_2 = FALSE \land$$

$$close\_comm\_failure = FALSE\} = (1 - p_1) \cdot (1 - p_2)^2.$$

Indeed, both barriers are closed only when the crossing controller received the close request and none of the barrier motors has failed. The calculation of the other two probabilities is slightly more complicated. Nevertheless, they can be straightforwardly obtained using the model safety invariants and basic rules for calculating probability. We omit the computation details due to a lack of space. The resulting probability of preservation of the safety property (1) is:

$$P_{SAF} = (1 - p_1) \cdot (1 - p_2)^2 +$$

$$(1 - p_4) \cdot \left(1 - (1 - p_1)^3 \cdot (p_2 \cdot p_3 + (1 - p_2) \cdot (1 - p_3))^2\right) -$$

$$(1 - p_1) \cdot (1 - p_2)^2 \cdot (1 - p_4) \cdot \left(1 - (1 - p_1)^2 \cdot (1 - p_3)^2\right).$$

Please note that P_{SAF} is defined as a function of probabilities of component failures, i.e., probabilities $p_1, \ldots, p_4$. Provided the numerical values of them are given, we can use the obtained formula to verify whether the system achieves the desired safety threshold.

5   Discussion

5.1   Related Work

Formal methods are extensively used for the development and verification of safety-critical systems. In particular, the B Method and Event-B are successfully being applied for formal development of railway systems [12,5]. A safety analysis of the formal model of a radio-based railway crossing controller has also been performed with the KIV theorem prover [16,15]. However, the approaches for integrating formal verification and quantitative assessment are still scarce.

Usually, quantitative analysis of safety relies on probabilistic model checking techniques. For instance, in [11], the authors demonstrate how the quantitative model checker PRISM [17] can be used to evaluate system dependability attributes. The work reported in [8] presents model-based probabilistic safety assessment based on generating PRISM specifications from Simulink diagrams annotated with failure logic. A method pFMEA (probabilistic Failure Modes and Effect Analysis) also relies on the PRISM model checker to conduct quantitative analysis of safety [9]. The approach integrates the failure behaviour into the system model described in continuous time Markov chains via failure injection. In [14] the authors propose a method for probabilistic model-based safety analysis for synchronous parallel systems. It has been shown that different types of failures, in particular per-time and per-demand, can be modelled and analysed using probabilistic model checking.

However, in general the methods based on model checking aim at safety evaluation of already developed systems. They extract a model eligible for probabilistic analysis and evaluate impact of various system parameters on its safety. In our approach, we aim at providing the designers with a safety-explicit development method. Indeed, safety analysis is essentially integrated into system development by refinement. It allows us to perform quantitative assessment of safety within proof-based verification of the system behaviour.

5.2   Conclusions

In this paper we have proposed an approach to integrating quantitative safety assessment into formal system development in Event-B. The main merit of our approach is that of merging logical (qualitative) reasoning about correctness of system behaviour with probabilistic (quantitative) analysis of its safety. An application of our approach allows the designers to obtain a probability of hazard occurrence as a function over probabilities of component failures.

Essentially, our approach sets the guidelines for safety-explicit development in Event-B. We have shown how to explicitly define safety properties at different levels of refinement. The refinement process has facilitated not only correctness-preserving model transformations but also establishes a logical link between safety conditions at different levels of abstraction. It leads to deriving a logical representation of hazardous conditions. An explicit modelling of probabilities of component failures has allowed us to calculate the likelihood of hazard occurrence. The B Method and Event-B are successfully and intensively used in the development of safety-critical systems, particularly in the railway domain. We believe that our approach provides the developers with a promising solution unifying formal verification and quantitative reasoning.

In our future work we are planning to further extend the proposed approach to enable probabilistic safety assessment at the architectural level.

Acknowledgments. This work is supported by IST FP7 DEPLOY Project. We also wish to thank the anonymous reviewers for their helpful comments.

References

1. Abrial, J.R.: Extending B without Changing it (for Developing Distributed Systems). In: Habiras, H. (ed.) First Conference on the B Method, pp. 169–190 (1996)
2. Abrial, J.R.: The B-Book: Assigning Programs to Meanings. Cambridge University Press, Cambridge (2005)
3. Abrial, J.R.: Modeling in Event-B. Cambridge University Press, Cambridge (2010)
4. Avizienis, A., Laprie, J.C., Randell, B.: Fundamental Concepts of Dependability, Research Report No 1145, LAAS-CNRS (2001)
5. Craigen, D., Gerhart, S., Ralson, T.: Case Study: Paris Metro Signaling System. IEEE Software, 32–35 (1994)
6. EU-project DEPLOY, http://www.deploy-project.eu/
7. Feller, W.: An Introduction to Probability Theory and its Applications, vol. 1. John Wiley & Sons, Chichester (1967)
8. Gomes, A., Mota, A., Sampaio, A., Ferri, F., Buzzi, J.: Systematic Model-Based Safety Assessment Via Probabilistic Model Checking. In: Margaria, T., Steffen, B. (eds.) ISoLA 2010. LNCS, vol. 6415, pp. 625–639. Springer, Heidelberg (2010)
9. Grunske, L., Colvin, R., Winter, K.: Probabilistic Model-Checking Support for FMEA. In: QEST 2007, International Conference on Quantitative Evaluation of Systems (2007)
10. Hallerstede, S., Hoang, T.S.: Qualitative Probabilistic Modelling in Event-B. In: Davies, J., Gibbons, J. (eds.) IFM 2007. LNCS, vol. 4591, pp. 293–312. Springer, Heidelberg (2007)
11. Kwiatkowska, M., Norman, G., Parker, D.: Controller Dependability Analysis by Probabilistic Model Checking. In: Control Engineering Practice, pp. 1427–1434 (2007)
12. Lecomte, T., Servat, T., Pouzancre, G.: Formal Methods in Safety-Critical Railway Systems. In: Brasilian Symposium on Formal Methods (2007)
13. McIver, A.K., Morgan, C.C.: Abstraction, Refinement and Proof for Probabilistic Systems. Springer, Heidelberg (2005)
14. Ortmeier, F., Güdemann, M.: Probabilistic Model-Based Safety Analysis. In: Workshop on Quantitative Aspects of Programming Languages, QAPL 2010. EPTCS, pp. 114–128 (2010)
15. Ortmeier, F., Reif, W., Schellhorn, G.: Formal Safety Analysis of a Radio-Based Railroad Crossing Using Deductive Cause-Consequence Analysis (DCCA). In: Dal Cin, M., Kaâniche, M., Pataricza, A. (eds.) EDCC 2005. LNCS, vol. 3463, pp. 210–224. Springer, Heidelberg (2005)
16. Ortmeier, F., Schellhorn, G.: Formal Fault Tree Analysis: Practical Experiences. In: International Workshop on Automated Verification of Critical Systems, AVoCS 2006. ENTCS, vol. 185, pp. 139–151. Elsevier, Amsterdam (2007)

17. PRISM – Probabilistic Symbolic Model Checker,
    http://www.prismmodelchecker.org/
18. Rigorous Open Development Environment for Complex Systems (RODIN): Deliverable D7, Event-B Language, http://rodin.cs.ncl.ac.uk/
19. Rigorous Open Development Environment for Complex Systems (RODIN): IST FP6 STREP project, http://rodin.cs.ncl.ac.uk/
20. RODIN. Event-B Platform, http://www.event-b.org/
21. Tarasyuk, A., Troubitsyna, E., Laibinis, L.: Quantitative Verification of System Safety in Event-B. Tech. Rep. 1010, Turku Centre for Computer Science (2011)
22. Tarasyuk, A., Troubitsyna, E., Laibinis, L.: Towards Probabilistic Modelling in Event-B. In: Méry, D., Merz, S. (eds.) IFM 2010. LNCS, vol. 6396, pp. 275–289. Springer, Heidelberg (2010)

Experience-Based Model Refinement

Didier Buchs, Steve Hostettler, and Alexis Marechal

Université de Genève, Centre Universitaire d'Informatique
7 route de Drize, 1227 Carouge, Suisse

Abstract. The resilience of a software system can be guaranteed, among other techniques, by model checking. In that setting, it consists in exploring every execution of the system to detect violations of resilience properties. One approach is to automatically transform the program into a model. To harness the system complexity and the state space explosion, designers usually abstract details of the studied system. However, abstracting too many details may dramatically impact the validity of the model checking. This includes details about the execution environment on which resilience properties are often based. This article sketches an iterative methodology to verify and refine the transformation. We introduce the concept of witness programs to reveal a set of behaviors that the transformation must preserve.

1 Introduction

The ubiquity of IT-systems makes their robustness increasingly important. Embedded systems such as medical devices, cars or airplanes require a very high safety level because human lives are involved. Obviously, resilience aspects should be integrated in the software engineering lifecycle. By resilience, we mean the ability of a software system to adapt itself to maintain an acceptable level of service in the presence of impairments [1]. Resilience can be assessed using different approaches at different phases of the system development. Testing is arguably the most well known approach and the most used in the industry. Although very effective, it suffers from only verifying a limited number of scenarios. Faults such as *race conditions* and *overflows* are hard to detect using testing. Such limitations can be partially solved using *model checking* [2].

Model checking differs from testing in that it explores all the possible behaviors of an abstracted version of the system. There are two main approaches to use this abstracted version of the system:

- the *model-to-code* approach that is the basic building block of the Model-Driven Engineering (MDE) approach [3]. It consists in designing a model, verifying it, and finally transforming it into runnable code [4]. The transformation is essentially a specialization as it adds information to the initial model (*i.e.*, the formal model).

E.A. Troubitsyna (Ed.): SERENE 2011, LNCS 6968, pp. 40–47, 2011.

– the *code-to-model* approach consists in transforming the code to a formal model. This approach is mainly used when dealing with legacy code. Mainstream tools such as IBM's Rational Software Architect [5] provide such reverse engineering capabilities for documentation purposes. The transformation is an abstraction as it hides or even removes information from the the code. For instance, let us also cite the Java PathFinder (JPF) model checker [6] that extracts a transition system from Java code and evaluates properties on it. JPF uses a set of mechanisms (*e.g.*, choice generators, partial order reduction, listeners, ...) to select the states to observe, thus building an abstract model of the execution.

In this paper we consider the second approach, that is to extract a formal model from a given code base. The modeling of a program and its execution environment can be automated [7,8]. When building the model, making abstractions is necessary because usual software may have trillions of possible executions. Usually, *code-to-model* approaches do not consider to modify the transformation. We argue that depending of the kind of property to verify, the user may apply different abstractions. Furthermore, every transformation induces implicit abstractions. For instance, the specific details of the CPU behaviors or the full range the data domains are usually abstracted. Identifying the relevant abstractions is a nontrivial task. Making too many abstractions may hide erroneous behaviors, while making not enough abstractions leads to intractable model checking. Thus, it is important to verify that the transformation does not hide more behaviors than expected.

This article presents a quick glance to a work in progress at our group. It aims to improve the quality of *code to model transformations* by semi-automatically verifying it. Section 2 illustrates the problem through an example. Section 3 sketches an iterative methodology to verify and refine the transformation. It introduces the concept of witness programs. Section 4 discusses some related work. Finally, Section 5 discusses the open tasks and concludes.

2   Motivating Example

This section presents a toy example to illustrate our approach. First we describe a failed attempt at detecting a bug in an implementation of the *singleton* design pattern using model checking. This demonstrates that making the wrong abstractions may impair bugs detection. Then we show how we could use this failure to improve model checking on further programs.

In object-oriented languages, the singleton's purpose is to create at most one instance of a given class during execution. Listing 1 presents a Java implementation of the singleton. Experienced Java programmers immediately see the problem in this code; without any synchronization, multiple instances of the `Singleton` class can be created. Let us see whether model checking allows us to find this bug.

```
class Singleton {
    private static Singleton instance = null;

    private Singleton() {};

    public static Singleton getInstance() {
        if (instance == null) {
            instance = new Singleton();
        }
        return instance;
    }
}
```

Listing 1. Broken implementation of the singleton

2.1   Model Transformation

Figure 1 presents the transformation result of the code from Listing 1 into Petri nets with inhibitor arcs. The singleton specification stipulates that at most one instance must be created during the execution of the program. To detect this, the central place in Figure 1 counts the amount of created singleton instances.

The getInstance method in Listing 1 start with an if. This is translated in the model as a branching from the place P1. Two events can occur, depending on the value of the instance variable (either null or not-null). This model uses the place instances to check this condition. If this place is empty, it means that no singleton has been created, and thus the instance variable from Listing 1 is null. Otherwise, it means that instance is not null as any created Singleton instance is immediately assigned to the variable. This is an over-approximation we make for clarity reasons.

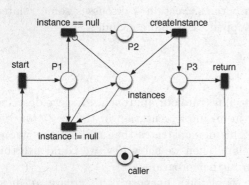

Fig. 1. Petri net model of Listing 1

As per Listing 1, if the instance variable is null, an instance of Singleton must be created. The transition createInstance represents the creation of a singleton. If the variable is not null no action is taken. Almost all places and transitions from Figure 1 correspond to a specific step or instruction in the singleton code. The only notable exception is the bottom place, which corresponds to the execution environment. The black token in place caller represents an object that calls the getInstance method and receives its result.

2.2   Failure and Correction of the Transformation

We can now check whether there is always at most one singleton in the model using a model checker. A cautious reader will immediately notice that no bug can be revealed using the model from Figure 1. The first time the method getSingleton is executed, the process creates a Singleton instance. On the following getSingleton calls, the process notices that a singleton was already created and thus does not create a new one. Nevertheless, as stated in the beginning of this section, the code from Listing 1 is broken. The bug in this code arises in a multithreaded execution.

The problem is that the modeling of the environment in Figure 1 (the place caller) only defines a single process executing the getInstance method. Observing the behavior of the code of Listing 1 in a multithreading environment requires to refine the transformation by adding several tokens in place caller. In this new model, the transition start may be fired multiple times, enabling the transition instance==null to be fired repeatedly. Hence, several processes may check concurrently that the instance variable is null prior to any instance creation. This leads to the creation of multiple instances. The *race condition* is now revealed. Clearly, this solution is not satisfactory for we need prior knowledge of the bug.

2.3   Lessons Learned

From the previous example arise two important remarks:

- Usually, most of the focus of the modeling activity is directed towards the code and not towards the environment. This example reveals a problem located in the transformation of the execution environment.
- Putting a single token in the bottom place is *not* an error. A model is an abstract representation of the system. We choose to abstract some behaviors in the model in order to be able to check it. Figure 1 was a good representation of Listing 1 for single threaded environments. The error was to abstract the multithreading, which might induce a bug.

In the singleton example, the model checking approach failed because the model did not take into account multithreading, an essential trait of the execution environment. This knowledge could be used to avoid repeating the same mistake in future checks. Any program that can be executed in a multithreaded environment (*i.e.*, almost every program nowadays) is subject to race conditions, like our singleton. For instance, the code in Listing 2, which represents a toy database that links String keys to java integers, tries to avoid duplicate keys. A *race condition* appears with multiple threads, just like for the *singleton*. A programmer may not know about the race condition bug, but he knows that his code will be executed in a multithreaded environment. For example, this knowledge may have been taken from the non-functional requirements of the software. If the programmer wants to use model checking to find errors, he should use a transformation that does not abstract multithreading.

```
class Database {
    private static Map map = new HashMap();

    public static void add(String key, Integer value) {
        if (!map.containsKey(key)) {
            map.put(key, value);
        }
    }
}
```

Listing 2. Broken implementation of a database

To verify this, the programmer can use the transformation on the singleton code of Listing 1 and perform model checking. If the singleton bug is revealed, there is a good chance that the transformation is correct. Otherwise, it probably means that the transformation abstracted the multithreading aspect. This transformation should be corrected before using it on the database code.

The singleton code of Listing 1 along with its bug description are used as a witness to show that a transformation preserves multithreading.

3   Proposed Methodology

Section 2 highlights the fact that the environment matters for model checking correctness. Resilience properties are often linked to the execution environment. Therefore, making too many abstractions may hide behaviors that are crucial to the verification of resilience properties. To circumvent this problem, one must verify that the transformation preserves the relevant details of the program and its environment. To that purpose, we propose to identify a set of witnesses that illustrate behaviors related to specific aspects of the environment. In this context, a witness is a small program that has a specific behavior under specific environment conditions. To be more precise, a witness is a tuple $\langle W_P, W_\Phi, W_T \rangle$ that consists in a program W_P, a property W_Φ to denote a given behavior, and a taxonomy label W_T that helps selecting the right witnesses.

For instance, the singleton code from Listing 1 is an example of a witness program (W_P). The corresponding property (W_Φ) denotes that at least two singletons can be instantiated. The corresponding taxonomy label (W_T) indicates that this behavior appears in a multithreaded environment.

One can check whether a *code to model* transformation preserves multithreading by verifying that its application on the witness code reveals the corresponding property. If the property does not hold on the transformed model we can conclude that multithreading is not preserved. With enough carefully selected witnesses like this one, it is possible to verify that a *code to model* transformation preserves a set of desired traits of the environment. This leads us to the methodology sketched in Figure 2.

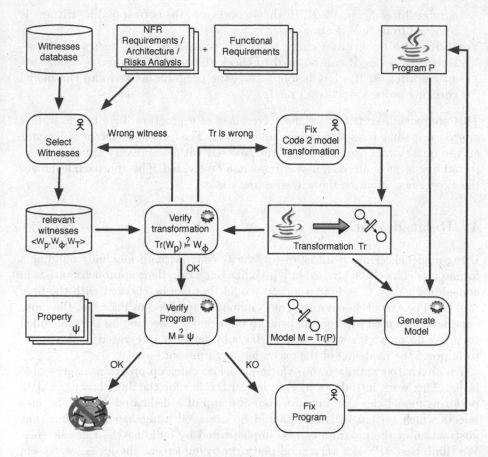

Fig. 2. The Approach

The methodology consists of three main steps:

Selecting the witnesses The standard software analysis documents (non-functional requirements, architecture, risks analysis, and the functional requirements) are used to define the characteristics of the execution environment. For instance, a web application that handles thousands of concurrent connections must obviously support multithreading. This human activity is called "Select the witnesses" in Figure 2 and produces a list of witnesses that are relevant to the software to verify.

Verify the code-to-model transformation The *code-to-model* transformation Tr is checked against the selected witnesses $\langle W_P, W_\Phi, W_T \rangle$. For a given witness, this amounts to transform the code of the witness using Tr to verify that the witness property holds $Tr(W_P) \models W_\Phi$. If the property does not hold, it means that either the *code-to-model* transformation is wrong, or the witness was mistakenly selected. In the first case, the transformation is refined (see Section 2.2). In the second case, the corresponding witness is removed from the selection.

Verifying the program Finally, when the transformation satisfies all the selected witnesses, it can be used to check the actual program P against its specification Ψ. If a bug is found then the program is fixed and the model checking process starts over again.

This approach does not prove the correctness of a program. Like in the testing approaches, some bugs may remain undetected. The level of correctness depends on the quality and exhaustivity of the selected witnesses. Every time an undetected bug is encountered a new witness can be created. The approach leverages the experience to enrich the witnesses database.

4   Related Work

Using model checking to automatically verify code, without manually building a formal model has been studied in [7]. Unlike our work, their approach consists in embedding native C code in a formal model that can be checked with the Spin tool [9]. The execution environment is manually modeled in Promela, the modeling language used by Spin, in order to allow model checking on the embedded code. As far as we know, our approach could perfectly integrated in this work, to improve the modeling of the execution environment.

An alternative approach to perform model checking on programs is presented in [6]. This work introduces JPF, a model checker for the Java language. JPF performs model checking on Java code, on top of a dedicated JVM. The user selects which system behaviors should be observed using *listeners*. Thus, any abstraction of the program can be implemented by building the right listener. We think that JPF is a very good platform to implement the ideas we present in the current article.

VerifyThis [10] is a global repository of known verification problems. Currently there are about 40 entries, some of them present a program and an undesired behavior. Thus, they are perfect candidates for the witnesses database.

5   Conclusion and Future Work

In this article, we show through an example that model checking programs with a *code to model* transformation highly depends on the quality of this transformation. It must preserve the relevant details including some environmental properties. To alleviate this problem, we sketch a methodology to verify that a transformation did not make undesired abstractions. This is particularly important for resilient properties, as they are usually heavily related to the execution environment. The transformation is refined through a set of witness programs harvested from the programmers experience.

The methodology described here is a work in progress with promising research perspectives and challenges:

- Guidelines for the refinement of the *code to model* transformation are needed.
- The VerifyThis database [10] could provide an initial load for the witnesses database. Means of enriching this database must be investigated.
- The integration of our approach in the software engineering cycle must be studied. The selection of witnesses may be based on standard software engineering documents.
- To fully comprehend the methodology of Figure 2, each step must be completely detailed and formalized.

These subjects will be the subject of our immediate future research. We plan to implement a proof of concept using JPF [6].

Acknowledgments. The authors would like to thank the editor and anonymous reviewers for their suggestions for improving the paper.

References

1. Di Marzo Serugendo, G., Fitzgerald, J., Romanovsky, A., Guelfi, N.: A metadata-based architectural model for dynamically resilient systems. In: Proceedings of the 2007 ACM Symposium on Applied Computing, SAC 2007, pp. 566–572. ACM, New York (2007)
2. Clarke, E.M., Emerson, E.A.: Design And Synthesis Of Synchronization Skeletons Using Branching Time Temporal Logic. In: Grumberg, O., Veith, H. (eds.) 25 Years of Model Checking. LNCS, vol. 5000, pp. 196–215. Springer, Heidelberg (2008)
3. Schmidt, D.C.: Model-driven engineering. IEEE Computer 39(2), 25–31 (2006)
4. Kraemer, F.A., Herrmann, P.: Automated encapsulation of uml activities for incremental development and verification. In: Schürr, A., Selic, B. (eds.) MODELS 2009. LNCS, vol. 5795, pp. 571–585. Springer, Heidelberg (2009)
5. IBM: Rational Software Architect for WebSphere Software (2010), http://www-01.ibm.com/software/awdtools/swarchitect/websphere/ (visited in June 2010)
6. Visser, W., Havelund, K., Brat, G., Park, S., Lerda, F.: Model checking programs. Automated Software Engineering 10, 203–232 (2003), doi:10.1023/A:1022920129859
7. Holzmann, G.J., Joshi, R., Groce, A.: Model driven code checking. Automated Software Engineering. 15, 283–297 (2008)
8. Di Marzo Serugendo, G., Guelfi, N., Romanovsky, A., Zorzo, A.F.: Formal development and validation of java dependable distributed systems. In: Proceedings of the 5th International Conference on Engineering of Complex Computer Systems, ICECCS 1999, pp. 98–108. IEEE Computer Society, Washington, DC, USA (1999)
9. Holzmann, G.: Spin model checker, the: primer and reference manual, 1st edn. Addison-Wesley Professional, Reading (2003)
10. Klebanov, V.: Verify This, http://verifythis.cost-ic0701.org/

Architecting Resilient Computing Systems: Overall Approach and Open Issues*

Miruna Stoicescu, Jean-Charles Fabre, and Matthieu Roy

CNRS ; LAAS ; 7 avenue du colonel Roche, F-31077 Toulouse , France
Université de Toulouse ; UPS, INSA, INP, ISAE ; UT1, UTM, LAAS

Abstract. Resilient systems are expected to continuously provide trustworthy services despite changes in the environment or in the requirements they must comply with. In this paper, we focus on a methodology
to provide adaptation mechanisms meant to ensure dependability while
coping with various modifications of applications and system context. To
this aim, we propose a representation of dependability-related attributes
that may evolve during the system's lifecycle, and show why this representation is useful to provide adaptation of dependability mechanisms at
runtime.

1 Introduction

One of the main challenges nowadays, as stated by IBM in The Vision of Autonomic Computing [12], is managing systems for which the total cost of ownership
is ever-increasing as they continuously evolve. The solution to this problem would
be for such systems to become autonomous, to a certain extent, and no longer
depend on humans for performing basic management tasks.

Autonomic Computing is also enticing for ubiquitous systems based on technologies such as Wireless Sensor Networks. The aim of Autonomic Computing
is described in [22] as addressing "today's concerns of complexity and total cost
of ownership while meeting tomorrow's needs for pervasive and ubiquitous computation and communication".

Our current work shares this vision while adding a fault tolerance axis. A
self-healing system is able to identify when its behaviour deviates from the expected one and to reconfigure in order to correct the deviation. We understand
a self-healing system as a context-aware fault tolerant system. To ensure a safe
adaptation, a validation step has to be added to this scheme, guaranteeing safety
during any reconfiguration. More precisely, we define the dynamic adaptation,
or self-healing, process as a *two-step permanent loop* consisting of a monitoring
service and an adaptation engine. The monitoring service is in charge of observing the system, measuring certain parameters and resource properties and
informing the adaptation engine. The latter must analyze the values, compare
the observed behaviour to the expected one, decide if an adaptation is needed,
choose a reconfiguration strategy and apply it.

---

* This work is supported by ANR, contract ANR-BLAN-SIMI10-LS-100618-6-01.

E.A. Troubitsyna (Ed.): SERENE 2011, LNCS 6968, pp. 48–62, 2011.

Our aim is to develop a method to define fault tolerant applications and to adapt them during the system's lifetime. The basic idea is to break them down into fine-grained components and to minimize the modification to be performed to update the fault tolerance mechanisms with respect to operational conditions.

This paper presents *on-going work* on a methodology for developing resilient applications for systems ranging from workstations to smart sensors. These applications are context-aware and must be able to adapt their fault tolerance mechanisms as a consequence of changes in their requirements and/or the environment. The rest of this paper is organized as follows: Section 2 gives our view on this matter and a representation of the problem, Section 3 thoroughly describes our approach, Section 4 presents our work plan, Section 5 presents a case study, Section 6 presents related works and finally, in Section 7 we give the concluding remarks.

2 Problem Statement

The evolution of systems may have an impact on dependability, i.e., various changes may invalidate some dependability properties and thus call for new solutions. In our work, we consider that any application is attached to one or several fault tolerance mechanisms (FTMs). Among the evolution scenarios that can occur during the operational life of a fault-tolerant application, we focus on those that may have an impact on the FTMs.

Beforehand, we define a frame of reference as a 3-axis space, each axis representing multiple variables, namely application assumptions, fault tolerance requirements, and system resources. A fault tolerant application may evolve in this space during the lifetime of a system due to:

- changes in the application assumptions which influence the choice of the fault tolerance mechanism,
- changes in the fault tolerance requirements,
- changes in the system resources (e.g., number of processors, memory resources, network bandwidth).

The essence of a *resilient application* [14] lies in the fact that when faced with any of these changes or with a combination of them, it must adapt itself while ensuring the observance of the dependability criteria. The 3-axis space gives an idea of the change model that we consider.

In Fig. 1, a cloud represents an application attached to a FTM. For a given application, each cloud represents a region in this 3-axis space. We view the evolution of a system during its operational life as a trajectory in a space characterized by the parameters which can cause the aforementioned evolution. The three evolution scenarios can be aggregated into this frame of reference for this space, the three axes being: the application assumptions (i.e., whether it is deterministic or not), the fault tolerance requirements (i.e., the fault model) and the current system resources.

It is the responsibility of the designer of the system to design the set of FTMs and foresee all possible evolutions and changes of the system, so that the

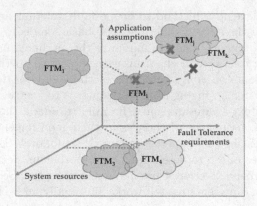

Fig. 1. Context-aware fault tolerant computing

trajectory it follows during its operational life is entirely covered by FTMs. We claim that a system's evolution can be represented in this vector space as shown in Fig. 1. The combination of these three types of information indicates the most convenient FTM to use. A solution in this space addresses a specific fault model, relies on some specific application assumptions and has a cost determined by a function with multiple resource parameters.

Given this frame of reference, we consider that each FTM covers a certain region of space as shown in Fig. 1. Obviously, the regions associated to different FTMs may partially overlap. To link the system's current state with this representation, a monitor keeps track of changes in terms of the three axes and allows placing the system status in a certain region of this space, which, by hypothesis, is covered by at least one FTM.

Once our view of the problem stated, a certain number of issues arise, such as: how do we associate FTMs with the combination of the three measures? how do we describe these FTMs in a way which allows us to change them dynamically? how do we actually perform the transition from one FTM to another? The following sections present our view on these matters.

3   Our Approach

The work reported in this paper is based on former work carried out on resilient computing at LAAS. We investigated the on-line adaptation of component-based FTMs (primary-bakup and semi-active replication) and applied it to an automatic control application (an inverse pendulum control system located on a cart). The modeling of FTMs using Petri nets was used to control their execution and to synchronize the adaptation, i.e. the change of mechanisms. An interesting point was the analysis and the definition of suitable adaptation states for performing the adaptation in a consistent manner. A full account of this work can be found in [8]. Another aspect of the work was the monitoring of timed

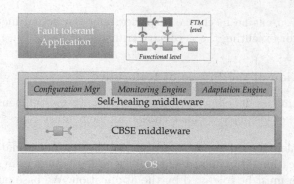

Fig. 2. The architecture of a reconfigurable fault tolerant application

properties of real-time applications, using Timed Automata. A monitor checking behavioural and timing properties was developed and implemented in Xenomai. Such a monitoring system was used to provide early error detection for fault tolerance strategies (see [9]). The approach proposed in this paper is based on the lessons learnt from these works and tries to define a whole development and runtime process to address the problem of dependable systems evolution.

Separation of concerns is a generally accepted idea for introducing FTMs in an application in a flexible way which allows subsequent modification and reuse. Based on the principle of behavioural reflection [16], we describe our approach using reflective component models (see Section 4.3). According to these principles, software architectures consist of two abstraction levels where the base level provides the required functionalities and the top level contains the FTM(s) [7]. As we target the adaptation of FTMs, we must manage the dynamics of the top level, which can have two causes:

- The application level remains unchanged but the FTM must be modified either because the fault model changes or because an event in the environment, such as resource availability, makes it unsuitable or, at least, non-optimal from a performance viewpoint.
- Changes in the top level are indirectly triggered by modifications in the application level which make the FTM unsuitable. In this case both levels are modified.

In order to achieve the adaptation of the FTMs in both scenarios, we build on the representation from the previous section and refine it in three steps.

Fig. 2 shows the "big picture" of the system architecture that we target. Based on a CBSE middleware, we develop a self-healing middleware populated with design patterns for fault tolerance and key services:

- a FTM repository in charge of storing the descriptions of FTMs in the form of component-based software architectures (see Section 4.2),
- a monitoring engine in charge of observing changes in the available system resources detailed in the next section,

- an adaptation engine in charge of performing the actual modifications on the FTMs for executing the necessary transitions.

3.1   Description of the Frame of Reference

The three parameters labeling our axes are actually multidimensional vectors but for the sake of visibility we have chosen an elementary representation.

- The fault tolerance requirements are part of the non-functional specifications of an application. Our main focus is on the fault model, i.e., the types of faults which must be tolerated by the application. We base our fault model classification on known types [1], e.g., physical faults, design faults, leading to value faults or crash faults only.
- The application assumptions group the characteristics of an application which have an impact on the choice of the FTM but are not the same as the functional specifications of the application. These characteristics include: whether the application is stateless or stateful, whether its state is accessible or not and whether the application is deterministic or not.
- The system resources axis groups information such as the number of available processors, the available bandwidth, the memory resources, ... Some of these variables can be precisely determined a priori (e.g., the number of processors) for a given FTM, while others are application-dependent (e.g., network bandwidth for replica synchronization). All these parameters must be continuously monitored.

3.2   Fault Tolerance Mechanisms

In order to place the FTMs in the previously described frame of reference, we must first be able to classify them using the given criteria. In our approach, there are three steps in the selection process for finding the most adequate FTM for an application: first, the fault model is identified, then the application assumptions and finally the currently available system resources. The same process will be applied for classifying them, resulting in a tree detailed in Section 4, Fig. 3.

3.3   System Evolution

The status of a fault-tolerant application represents a point in the space given by our frame of reference. The application being fault-tolerant, this point must be in a region covered by a certain mechanism. A change (in terms of one of the three axes) is equivalent to a new status of the application, therefore a new point. As our purpose is to guarantee dependability when facing changes, the application must always be "accompanied" by an adequate FTM on its evolutionary trajectory. Therefore, we must provide a FTM encompassing a region which contains the new point. We must design and build our FTMs in view of such transitions and adaptations. The "distance" (as will be defined later on in Section 4.4) between two mechanisms as represented in our frame of reference

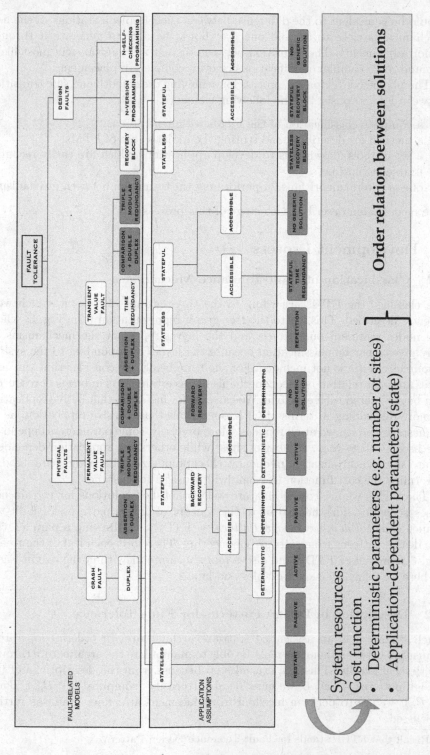

Fig. 3. Classification of fault tolerance mechanisms

should be equivalent to the difference between their implementations. If the new mechanism is close to the old one, we should be able to generate it through minor adjustments. The transition between distant mechanisms will most likely demand more complex modifications or even complete replacement.

The planned development process for achieving the smooth and safe transition between FTMs consists of the following elements:

- a complete classification of the FTMs we intend to use;
- a method for describing and storing the architecture of FTMs;
- a set of tools allowing us to develop applications which are easily reconfigurable at runtime;
- one or several algorithms for performing the transition between mechanisms.

The next section describes each stage of this process.

4   Development Process

4.1   Classification of Fault Tolerance Mechanisms

We classified the FTMs according to the three groups of criteria, which were already described. This led to the tree-graph representation from Fig. 3, where the nodes represent our criteria and the leaves represent the mechanisms. It only gives a partial classification because the layer corresponding to the system resources criteria is not included. For the time being, we consider that this axis gives an order relation between mechanisms as certain costs in terms of resources can be associated with each one. We can say, for instance, that active replication is more demanding in terms of CPU operations than passive replication, as all requests are processed at all replicas. More precisely, some parameters depend on the used strategy (e.g., number of sites) while others are application-dependent (state size, synchronization frequency, checkpoint size), the latter being used to determine the cost function (for bandwidth usage, for instance).

The FTMs we are considering are well-established solutions for certain use-cases/scenarios. This has led us to the concept of design patterns for fault tolerance (*Fault Tolerance Design Pattern* — FTDP), by drawing a parallel with the design patterns from software engineering. Therefore, each leaf of the tree in Fig. 3 represents a FTDP and also a more accurate view than the one in Fig. 1 as here we refine the axes variables and labels.

4.2   Description of Design Patterns for Fault Tolerance

Each FTDP[1] has an associated software architecture. In order to operate a transition between them, we must be able to manipulate the architecture through its description file. This is a crucial step because from the descriptions of two design patterns A and B, we must be able to easily "compute" $A \cap B$, A/B and $A \backslash B$, if we are to perform mechanism replacement at a finer grain, as further discussed.

---

[1] Recall that FTDP stands for Fault Tolerance Design Pattern.

A decision must be taken regarding what needs to be stored off-line, before the application executes. One idea would be to describe the architecture of all the FTDPs we intend to use and store these files. Another idea would be to group closely related design patterns, choose the most "representative" one for each class and store its description as well as describe the strategy for passing to another one.

An ADL, such as ACME [11], can be used for describing these architectures. ADLs facilitate the construction of models in which the architecture of a system is described as a composition of components and connectors. Some of them also include properties and constraints which can be placed on those entities. The use of an ADL enables us to reason about a software architecture during the specification stage. In order to maintain this reasoning consistency at runtime, we will use a component-based middleware for implementing the system. This stage is detailed in the next subsection.

4.3 Runtime Support and Reconfiguration

A design principle that is commonly accepted in the area of reconfigurable systems is the use of component-based technologies (CBSE) [6] for developing the management framework. CBSE middleware provide means to develop and run software as a graph of interconnected components. Applications built using a component-based middleware consist of components and connectors and can be modified at runtime by using the methods provided by the middleware for stopping, unbinding, binding and starting components. In this paper, this technology will be used to implement FTDPs whereas the application will be seen as a black box.

As there are many available platforms, during this project stage we will focus on finding a component-based middleware tailored to our needs. It should provide runtime reconfiguration facilities, have a small memory footprint, suiting target systems such as smart sensors with limited resources and provide the right level of design complexity for our problem. A first step in this direction will be to define a minimal set of instructions we need for our reconfigurations, detailed in the next section.

4.4 Transitions between Mechanisms

A crucial point concerns the granularity of the reconfiguration. There are two obvious approaches for replacing a FTM with another one:

- an atomic approach: the old mechanism is replaced by the new one,
- a differential approach: transition operates at a finer level by adding/removing components corresponding to the difference between the two mechanisms.

The *distance* between the two mechanisms plays an essential part in the choice of the approach. The function for computing this measure is, intuitively, the sum of three basic distances, namely application assumptions, fault model complexity and number of components to be changed. Regarding application assumptions, a stateless deterministic application has an associated weight lower than a

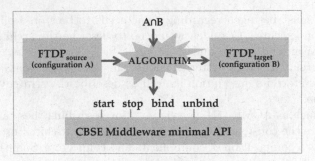

Fig. 4. Transition between FTMs using a minimal CBSE support

stateful non-deterministic application. Regarding fault model complexity, design patterns that tolerate crash faults have a lower complexity than design pattern that tolerate design faults. However, the most influencing parameter of this distance is the number of components to be changed between two component-based implementations of fault tolerance mechanisms.

The transitions between mechanisms will most likely be described through state-machines. The algorithms behind the transitions could lead to design patterns for system evolution.

The transition algorithm is executed by the adaptation engine represented in Fig. 2 and based on a minimal interface enabling components to be manipulated at runtime (see Fig. 4). Our final aim is to define the minimal interface required to install a component at runtime within a software configuration. Such an interface consists of at least four operations to manipulate components and bindings/connections: stop, start, bind, unbind. It also needs customized, user-defined methods like setState and getState operations for stateful objects. The stop operation is certainly more subtle than the others: its semantics is "stop after termination of all operations in progress". This means that all input requests are queued while requests already in progress terminate.

Moreover, any algorithm that performs an online adaptation of a mechanism based on a CBSE middleware must rely on the notion of *transaction*: when a transition from a design pattern to another one is triggered, the different components that have to be replaced must be changed on an *all-or-nothing* approach, i.e., the set of basic replacements must be encapsulated in a global transaction — when all components are modified, then the modification can be committed; in case of a single failure, the whole transaction is rolled back.

5   Case Study

The detailed classification of our FTMs given in Fig. 3 is the basis for their adaptation[2]. Any final FTDP is associated to a software architecture, i.e. the

---

[2] Notice that the rightmost leaf indicates that an ad-hoc solution must be defined, relying on a roll-forward recovery approach, i.e., defining intermediate valid states in the computation from which the application can restart.

FTM, ideally a component graph. Our final aim is to examine various transitions of the change model. This will consist in choosing a starting point, a destination point and several intermediate points, which all represent system state regions, associating the most adequate FTM to each one, and develop algorithms to perform safe transitions between mechanisms.

The FTDPs, more precisely their architectural description (component view) is the basis for a first analysis of the adaptation. Starting from one mechanism, called $FTDP_{source}$, the work consists in analyzing the changes to be performed to reach another mechanism, called $FTDP_{target}$. For each transition between $FTDP_{source}$ and $FTDP_{target}$, an algorithm has to be defined off-line to perform the change. This algorithm thus takes two configurations derived from the design descriptions given as inputs and performs the operation using the API of a CBSE middleware providing control over the components' lifecycle and bindings.

As an illustration, let us consider two variants of duplex strategy, namely a passive replication design pattern as a $FTDP_{source}$, and an active replication design pattern as a $FTDP_{target}$ (see Fig. 5). Passive replication is a checkpointing-based fault tolerance strategy for crash faults, with only one replica processing requests (PBR, *Primary Backup Replication* model). Active replication, in this case, is a semi-active fault tolerance strategy for crash faults, with both replicas processing requests and only one delivering output messages (LFR, *Leader-Follower Replication* model) (see [17]). Here, we consider simplified implementations of both strategies we developed in UML using RSA (*IBM Rational Software Architect*) tool. The complete design has been created and the passive replication strategy is described by the class diagram shown in Fig. 6. We assume that objects can be mapped to components to enable easy manipulation at runtime. In the design provided here, we consider the following objects mapped to components: Proxy-Server, Communication, Server (ServerPBR), DuplexProtocol, RemoteCall and the Client.

The proposed simplified design perfectly illustrates our method, i.e., it entails minimal modifications to execute a transition to an active replication strategy. The distance being small in this case, the transition corresponds to replacing the

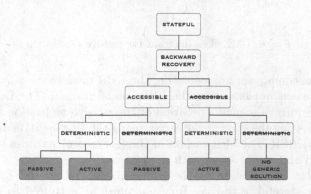

Fig. 5. DUPLEX/STATEFUL subtree

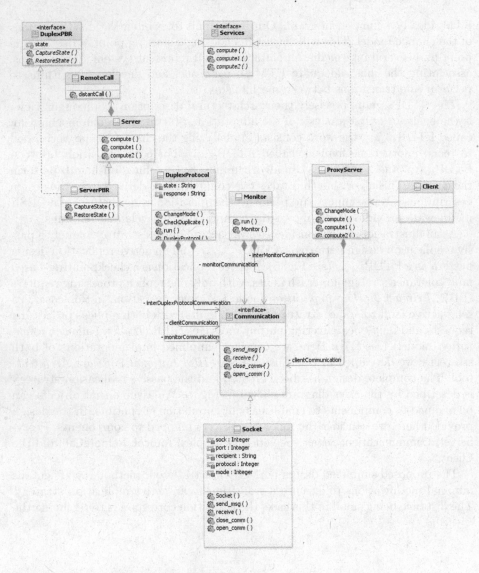

Fig. 6. UML class diagram for passive replication

DuplexProtocol component implementing the PBR model with a DuplexProtocol component variant implementing the LFR model instead. The DuplexProtocol component performs the main operation of the passive replication strategy (run method), by getting the state of the server using the CaptureState method inherited from ServerPBR. The LFR variant of DuplexProtocol activates the requested server method and synchronizes with the follower.

With respect to the state of the protocol, i.e., either the DuplexProtocol is stateless or stateful, a transfer function must be activated to initialize the new DuplexProtocol component. In our case, to simplify implementation, we considered it stateless. Notice that this assumption greatly simplifies switching, as explained in [8].

6 Related Works

As already mentioned in Section 1, one of our first areas of interest was Autonomic Computing (AC), more precisely self-healing systems, i.e., systems which are able to repair themselves. However, as [22] also points out, many of the "hot" issues within AC have been at the core of other disciplines, e.g., fault tolerant computing and artificial intelligence for a long time. The novelty lies in the "holistic aim" of regrouping all relevant research areas in this common project. Focusing on the intersection between AC and fault tolerance computing, which is our main research axis, the same author, in [23], reaches the conclusion that dependability and fault tolerance are not only "specifically aligned to the self-healing facet" of AC but, on a closer view, "all facets of Autonomic Computing are concerned with dependability" (i.e., self-configuration, self-optimization and self-protection as well).

A complex example of (re)configurable framework which also touches the areas of ubiquitous systems, AC and fault tolerant computing is Gaia [19], which builds on the core concept of *active spaces*. Heterogeneous devices, such as PCs, sensors, smart-phones, blend in the environment and interact with the user through a uniform interface. A method of adding an autonomic dimension to this framework is the use of a planner from the field of artificial intelligence, as presented in [18], which has the role of creating a path between the current state and a goal state given by a user or by a developer. This planner can be considered analogous to our transition algorithm. In [3], the authors add a fault tolerance dimension to Gaia by emphasizing the importance of dependability in pervasive systems i.e. systems which interact closely with the users, possibly even in a healthcare context. They only endow Gaia with the property of tolerating fail-stop faults. Although this framework tries to cover issues from many areas including dependable computing, the proposed solution appears to be closely linked to an underlying operating system called 2K, the case studies focus only on managing media presentations with a basic support for fault tolerance. RUNES (Reconfigurable Ubiquitous Networked Embedded Systems) [5] is also a project addressing several of the areas mentioned above: it is an easily tailorable middleware component framework for resource-constrained systems (e.g., sensor networks) which aims at the reconfiguration of heterogeneous systems. The scenario they address in [4] involves fire management in a road tunnel that is instrumented with sensors and actuators communicating with devices such as PDAs belonging to firemen.

In the area of reconfigurable software architectures, RAINBOW [11] builds on the use of architectural models for problem diagnosis and repair. The proposed framework includes a monitoring layer composed of two types of entities, namely probes which gather basic data from the running system and gauges which perform computations on the data in order to obtain measures of the system properties. An architecture manager is in charge of maintaing the architectural model at runtime and of verifying that the constraints on the system elements are maintained. The project is very complex as it englobes a very expressive ADL called ACME [10], already cited, a system in charge of verifying constraints, called ARMANI, a library of gauges, etc. The idea of placing an ADL on top of a CBSE middleware, as we intend to do, is the topic of [2], in which the authors describe their experience in associating an enriched version of ACME with the OpenCOM middleware for providing programmed and ad-hoc changes at runtime while maintaining certain constraints. Drawing the parallel between the entities described by the ADL and those provided and manipulated by the CBSE will also be an important step in our research.

Although there are a lot of available component-based platforms, not all of them are equally suitable for runtime reconfiguration and for dealing with self-healing. In [21], the authors identify the necessary runtime abstractions that a component model must include in order to efficiently support an autonomic adaptation service. What we need is not a component model which merely allows runtime reconfiguration but one which is designed for it. An example is [20] which builds, as RUNES, on the idea of a basic runtime kernel and a variety of additional services in the form of modules which can be plugged in. A fertile source of inspiration are the projects aimed at providing reconfiguration/adaptation properties to wireless sensor networks such as the works described in [24], [25] and [26] which present a middleware for wireless sensor networks and a component model enabling fine-grained adaptation.

The idea of creating design patterns for fault tolerance based on meta-level architectures and fault tolerance mechanisms is also described in [15]. [13] presents reflective design patterns implementing the separation of concerns as well as a hierarchy of design patterns in the form of a tree-graph where each level acts as a refinement step.

7   Conclusion and Perspectives

In this paper, we presented our approach to design and implement *resilient systems*, i.e., systems meant to cope with continuous evolution while guaranteeing dependability. Hence, our work lies at the intersection of three research domains, namely Fault Tolerance, Autonomic Computing and Ubiquitous Systems.

To that aim, we propose to represent various dependability-related attributes of the system in a three-axes space. Interestingly, this space allows us to link application and dependability-related context with fault tolerance strategies. More precisely, we classified classical fault tolerance patterns with regards to this space, and showed how every fault tolerance pattern defines a region in this space.

In this formalism, the evolution of the system can be represented as a path in this space, and evolution induces modifications of fault tolerance strategies, as derived from our classification of fault tolerance patterns.

To illustrate our method, we show, on a simple active to passive replication evolution, how we can provide an algorithmic solution of this evolution, based on a simplified UML description of such replication strategies.

To go further this simple example, we are currently investigating the possibility to automatically link fault tolerance design patterns (such as the one provided in UML) to component-based technologies (CBSE) for developing the management framework. Using such approach will allow us to define precise algorithms for adaptation, and will ease the development of a generic framework for adaptation of dependable systems.

References

1. Avizienis, A., Laprie, J.-C., Randell, B., Landwehr, C.: Basic concepts and taxonomy of dependable and secure computing. IEEE Trans. Dependable Secur. Comput. 1, 11–33 (2004)
2. Batista, T., Joolia, A., Coulson, G.: Managing dynamic reconfiguration in component-based systems. Software Architecture, 1–17 (2005)
3. Chetan, S., Ranganathan, A., Campbell, R.: Towards fault tolerance pervasive computing. IEEE Technology and Society Magazine 24(1), 38–44 (2005)
4. Costa, P., Coulson, G., Gold, R., Lad, M., Mascolo, C., Mottola, L., Picco, G.P., Sivaharan, T., Weerasinghe, N., Zachariadis, S.: The RUNES middleware for networked embedded systems and its application in a disaster management scenario (2007)
5. Costa, P., Coulson, G., Mascolo, C., Mottola, L., Picco, G.P., Zachariadis, S.: Reconfigurable component-based middleware for networked embedded systems. International Journal of Wireless Information Networks 14(2), 149–162 (2007)
6. Coulson, G., Blair, G., Grace, P., Taiani, F., Joolia, A., Lee, K., Ueyama, J., Sivaharan, T.: A generic component model for building systems software. ACM Transactions on Computer Systems (TOCS) 26(1), 1–42 (2008)
7. Fabre, J.-C.: Architecting dependable systems using reflective computing: Lessons learnt and some challenges. In: Dehne, F., Gavrilova, M., Sack, J.-R., Tóth, C.D. (eds.) WADS 2009. LNCS, vol. 5664, pp. 273–296. Springer, Heidelberg (2009)
8. Fabre, J.C., Killijian, M.O., Pareaud, T.: Towards On-line Adaptation of Fault Tolerance Mechanisms. In: EDCC, pp. 45–54. IEEE, Los Alamitos (2010)
9. Fabre, J.C., Robert, T., Roy, M.: Early Error Detection for Fault Tolerance Strategies. In: RTNS. IEEE, Los Alamitos (2010)
10. Garlan, D., Monroe, R.T., Wile, D.: Acme: Architectural description of component-based systems. Cambridge University Press, Cambridge (2000)
11. Garlan, D., Schmerl, B.: Model-based adaptation for self-healing systems. In: Proceedings of the First Workshop on Self-Healing Systems, WOSS 2002, pp. 27–32. ACM, New York (2002)
12. Kephart, J.O., Chess, D.M.: The vision of autonomic computing. Computer 36(1), 41–50 (2003)
13. Lamour, L., Cecília, F., Rubira, M.F.: Reflective Design Patterns to Implement Fault Tolerance
14. Laprie, J.-C.: From dependability to resilience. In: DSN, Anchorage, AK, USA, vol. 8, pp. G8–G9 (2008)
15. Lisbôa, M.L.B.: A new trend on the development of fault-tolerant applications: software meta-level architectures. J. of the Brazilian Computer Society 4(2) (1997)
16. Maes, P.: Concepts and experiments in computational reflection. ACM Sigplan Notices 22(12), 147–155 (1987)

17. Powell, D.: Distributed fault tolerance Lessons learnt from delta-4. Hardware and Software Architectures for Fault Tolerance, 199–217 (1994)
18. Ranganathan, A., Campbell, R.H.: Autonomic pervasive computing based on planning (2004)
19. Román, M., Hess, C., Cerqueira, R., Ranganathan, A., Campbell, R.H., Nahrstedt, K.: A middleware infrastructure for active spaces. IEEE Pervasive Computing, 74–83 (2002)
20. Romero, D., Hermosillo, G., Taherkordi, A., Nzekwa, R., Rouvoy, R., Eliassen, F.: The DigiHome Service-Oriented Platform. Software: Practice and Experience (2011)
21. Sicard, S., Boyer, F., De Palma, N.: Using components for architecture-based management: the self-repair case. In: Proceedings of the 30th International Conference on Software Engineering, pp. 101–110. ACM, New York (2008)
22. Sterritt, R.: Autonomic computing. Innovations in Systems and Software Engineering 1(1), 79–88 (2005)
23. Sterritt, R., Bustard, D.: Autonomic Computing-a means of achieving dependability? In: Proc. IEEE Engineering of Computer-Based Systems (2003)
24. Taherkordi, A., Eliassen, F., Rouvoy, R., Le-Trung, Q.: ReWiSe: A New Component Model for Lightweight Software Reconfiguration in Wireless Sensor Networks. In: Meersman, R., Tari, Z., Herrero, P. (eds.) OTM-WS 2008. LNCS, vol. 5333, pp. 415–425. Springer, Heidelberg (2008)
25. Taherkordi, A., Le-Trung, Q., Rouvoy, R., Eliassen, F.: WiSeKit: A Distributed Middleware to Support Application-Level Adaptation in Sensor Networks. In: Senivongse, T., Oliveira, R. (eds.) DAIS 2009. LNCS, vol. 5523, pp. 44–58. Springer, Heidelberg (2009)
26. Taherkordi, A., Rouvoy, R., Le-Trung, Q., Eliassen, F.: Supporting lightweight adaptations in context-aware wireless sensor networks. In: Workshop on Context-Aware Middleware and Services/COMSWARE, pp. 43–48. ACM, New York (2009)

Supporting Architectural Design Decisions Evolution through Model Driven Engineering

Ivano Malavolta[1], Henry Muccini[1], and V. Smrithi Rekha[1,2]

[1] University of L'Aquila, Dipartimento di Informatica, Italy
[2] Department of Computer Science and Engineering, Amrita Vishwa Vidyapeetham, India
{ivano.malavolta,henry.muccini}@univaq.it,
smrithirekha@gmail.com

Abstract. Architectural design decisions (i.e., those decisions made when architecting software systems) are considered an essential piece of knowledge to be carefully documented and maintained. As any other artifact, architectural design decisions may evolve, having an impact on other design decisions, or on related artifacts (like requirements and architectural elements). It is therefore important to document and analyze the impact of an evolving decision on other related decisions or artifacts. In this work we propose an approach based on a notation-independent metamodel that becomes a means for systematically defining traceability links, enabling inter-decision and extra-decision evolution impact analysis. The purpose of such an analysis is to check the presence of inconsistencies that may occur during evolution. An Eclipse plugin has been realized to implement the approach.

1 Introduction

A new perspective on Software Architecture (SA) views an SA as a composition of a set of design decisions [1], taken to resolve architectural design issues. In this light, and as currently recognized by software architects, architectural design decisions (ADDs) have to be considered first class entities when architecting software systems [2,1,3]. Therefore, they have to be explicitly documented and maintained. Since ADDs strongly depend on elicited requirements and highly contribute to shape the architecture of complex software systems, it is highly recommended to support ADDs evolution and to explicitly document and analyze the impact of design decisions' changes to other decisions and to related artifacts [4,5,6]. While a few approaches and tools have been proposed for documenting and managing ADDs evolution, they are typically tied to some specific architecture description language or ADD notation, and they only partially document the impact of a decision change to other decisions and related artifacts.

In this paper we propose an approach to support ADD evolution. It provides a generic, ADD notation-independent approach that starting from a model based specification of ADDs, enables to identify the impact of a changing design decision on other ADDs (referred as *inter-decisions* analysis) and the impact of the same on requirements and architectural descriptions (named *extra-decisions* analysis).

Our contribution is mainly in the following lines: *a*) We have proposed a metamodel for evolving ADDs. The metamodel is generic and flexible enough to manage requirements and architectures described in any notation; *b*) the proposed metamodel enables

E.A. Troubitsyna (Ed.): SERENE 2011, LNCS 6968, pp. 63–77, 2011.

explicit representation of relationships among ADDs *c*) the metamodel enables bi-directional traceability links between ADDs, requirements and architectural elements which will help in analyzing the impact of evolution on the corresponding artifacts; *d*) we present an approach that enables inter-decision and extra-decision validation to check the presence of inconsistencies that may occur during evolution.

The structure of the paper is as follows. Section 2 provides a background on SA evolution and highlights the importance of explicitly supporting ADD evolution and motivates our work. Section 3 describes our approach. Section 4 presents a prototype implementation of what is presented before. Section 5 summarizes related works, while Section 6 concludes the paper.

2   Background and Motivation

All software systems are subject to constant evolution, which is often triggered by external stimuli. As a result, all related software artifacts must evolve to handle foreseen or unforeseen changes [7]. By looking for example at the Rational Unified Process (RUP) [8], we observe that the architecting process is spread over several iterations and each iteration results in an *executable architectural prototype*. This results in an evolving architecture, being refined through many incremental iterations. Many times, requirements and architectural design co-evolve owing to exploration, negotiation and decision-making [9]. Since ADDs are related to each other in different ways and are related to other artifacts (e.g., ADDs directly trace to requirements *upstream* and to design and implementation *downstream* [2]), such an incremental refinement process requires to constantly add or modify decisions, that will in turn impact other decisions, requirements and architectural components. Practically, architects face difficulties when managing evolution: if the relationships among different decisions and the mapping between ADDs, requirements and architectural artifacts are not clear, changes to one decision may conflict with other decisions, bringing the system to an inconsistent state.

In order to partly cope up with such scenarios, several tools and approaches are available to capture ADDs and rationale (Archium [1] and ADDSS [10], for example). Most currently available tools capture the following concepts: *issues* or *problems*, *alternatives* or *options*, *assumptions*, *constraints*, *design decisions*, *rationale*, and relationships among ADDs. Though these tools and methods effectively capture architectural knowledge, very few tools have been designed to account for evolution of architectural decisions. The lack of support or availability of only minimal support for evolution by current tools and Architecture Description Languages has been discussed in [6,7,11].

3   Supporting Design Decisions Evolution

This section presents our approach to support design decisions evolution analysis through bidirectional traceability between decisions and all the involved artifacts.

In this respect, it is fundamental to analyze how design decisions evolve, and the actual impact of a design decision evolution on other related artifacts like requirements specifications and software architecture descriptions.

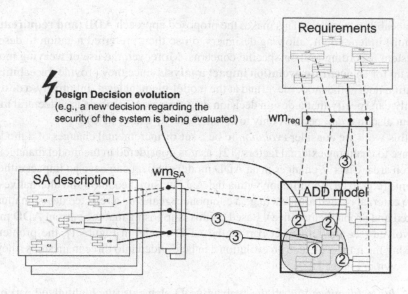

Fig. 1. Overview of the proposed approach

Few factors that may impact the evolution of ADD include:

1. Natural evolution that happens when the system is being developed incrementally
2. Evolution of requirements during the development of the system
3. Once the system has been deployed, Addition/Deletion/Changes to existing:
   (a) Requirements
   (b) Issues
   (c) Alternatives (e.g., technology alternatives) for an issue under consideration
   (d) Stakeholder concerns

These stimuli may directly impact the ADD model as well as its related artifacts. Figure 1 gives an overview of the approach we propose to support architectural design decisions evolution.

The **ADD model** is the most important artifact since it represents all the design decisions identified during the design process; it also classifies design decisions depending on their type and their status, and keeps track of the various relationships between them (Section 3.1 gives an overview on how to specify design decisions). Design decisions are linked to (i) **Requirements specification** and (ii) a set of **SA descriptions** resulting from the chosen design decisions (only one SA description is chosen, all the others are alternatives).

Traceability is realized as a set of tracing links from requirements to ADDs and from ADDs to SA descriptions (see the dotted lines in Figure 1). Tracing links are contained into special models (wm_{req} and wm_{SA} in Figure 1) called *weaving models*; they will be described in detail in Section 3.2. A peculiar aspect of our approach is that tracing links are represented as models; this aspect is perfectly in line with MDE and represents an interesting added value for our approach. In particular, the use of weaving models allows designers to keep the requirements and architectural models free from any traceability metadata. This makes the proposed approach **ADL (and requirements notation) independent**, allowing designers to use their preferred notation to describe the system according to their specific concerns. Moreover, the use of weaving models opens up for an **accurate evolution impact analysis** since they provide traceability information both at the model-level and at the model-element level. Intuitively, a designer not only can specify that a design decision dd_x pertains to a whole architectural model am, but also that dd_x pertains only to components c_1 and c_2 in am.

In this work we consider *evolution* to be a set of incremental changes of a model in response to changing external factors [12], as it is considered in the model management research area. Clearly, changes in an ADD model element can have an impact either on elements (e.g., ADDs conflicting with a deleted decision become valid alternatives), or also to external related artifacts (e.g., a component realizing a rejected decision must be checked and probably modified). Based on this, supposing that the current ADD model has evolved (see the left-lower part of the ADD model in Figure 1), the problem of analysing the impact of such an evolution can be divided into three main steps (they are also shown in Figure 1):

1. *evolution identification*: all the evolved ADD elements (the highlighted part of the ADD model in Figure 1) must be timely identified and must be presented to the designer;
2. *inter-decisions analysis*: the portion of ADD model which depends on the initial evolved ADD elements identified and their consistency must be checked;
3. *extra-decisions analysis*: all requirements and architectural elements depending on the evolved ADD elements are identified and checked against consistency.

It is important to check the ADD model and its related artifacts after evolution since evolution may may introduce contradictions between ADDs (or partially defined elements, and so on) that must be carefully checked by the tool and solved by the designer (this aspect will be clarified in section 3.3).

Next sections describe in detail how our proposed approach allows designers (i) to clearly specify design decisions and their relationships between each other (see Section 3.1), (ii) to trace design decisions towards other system life-cycle artifacts (see Section 3.2), and (iii) to estimate the impact of design decisions evolution (see Section 3.3).

3.1   ADDMM: A Metamodel for Supporting Design Decisions Evolution

In the proposed approach design decisions are represented within a Design Decision(DD) model. A DD model represents all the design decisions and keeps track of all the design reasoning history of the system. For the sake of generality, instead of being coupled to a specific notation for ADDs (and its corresponding tool), we propose $ADDMM$: a notation- and tool-independent metamodel for representing design decisions and their evolution.

The $ADDMM$ metamodel has been defined by analyzing the current state of the art ADD representation ([2], [13], [14] to include a few). We designed $ADDMM$ so that it (i) presents the most common aspects and relationships of design decisions, and (ii) contains concepts to define the evolution of an ADD (like its state, history and scope). It

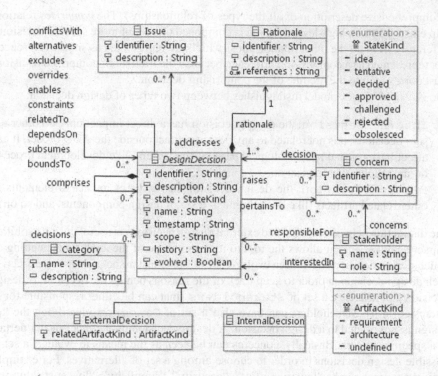

Fig. 2. ADDMM: A metamodel for evolving design decisions

is important to note that, even if we customized the proposed approach to $ADDMM$, it is possible to adapt the proposed approach to work with other metamodels for ADDs (the visualization and traceability parts of the approach remain unchanged, whereas the evolution analysis part must be adapted to the new metamodel since it directly refers to elements in $ADDMM$).

Figure 2 graphically shows the metamodel for design decisions models. Naturally, DesignDecision is the the main concept of the metamodel. It contains attributes like *description*, *state* that represents the current state of the decision (e.g., idea, tentative, decided, rejected, etc.), *timestamp* that specifies when the design decision has been created, *history* that keeps track of the evolution history of the decision (mainly, author, timestamp and status of each past version of the decision, and so on); it should be noted that the *history* attribute is conceived as a means for software architects to quickly access information about the evolution of a design decision over time, other mechanisms are used to manage different (evolved) versions of a design decision model (see Section 3.3). The *evolved* attribute is used as a marker to identify which design decisions have been subject to an evolution, it will be used during the evolution impact analysis, which is in charge of managing traceability between evolutions of architectural design decisions models (see Section 3.3).

Each design decision can be related to some other design decisions. Examples of relationships include: *constrains*, *enables*, *conflictsWith*, and so on (please refer to [2] for a comprehensive description of all the types of relationships). The *comprises* relationship exists when an high-level decision is composed of a set of more specific decisions; differently from all the other relationships (which are represented as simple reference), it is represented as a composition reference because the lifetime of comprised decisions must correspond to the lifetime of the comprising decision.

The $ADDMM$ metamodel distinguishes between two types of design decision:

1. `InternalDecision`: the design decision has a direct impact only on other design decisions; it is not related to any design artifact outside the ADD model. It can affect business and methodological parts of the system, like development process, organization issues, and so on.
2. `ExternalDecision`: the design decision has a direct impact to elements in other related artifacts (like a requirement, a set of design components, and so on).

The distinction between internal and external design decisions is useful during evolution impact analysis since it allows the tool to tune the scope of the analysis depending on which design decisions have been evolved. Designers must associate a `Rationale` to each design decision in order to keep track of the reasons behind the decision. A design decision can reference a set of `Stakeholders` (that can be either responsible for or interested in it). Stakeholders can have also a set of `Concerns` identifying the key design issues related to a design decision. A design decision may either raise or pertain to a specific concern. Basically, concerns can be seen as the means to evaluate a set of possible design decisions in order to choose among a set of alternatives. For example, typical concerns can be "low user effort", "security", "simplicity", "low cost solution", and so on. In order to keep design decisions cognitively manageable and well scoped, they can be organized into `categories`.

Providing a concrete syntax for ADD models is out of the scope of this work, so we have defined a simplified graph-based representation that renders ADDs as nodes of the graph and their relationships as edges between nodes. An example of this pseudo-notation is shown in Section 4.2.

3.2   Tracing Design Decisions to/from other Artifacts

As introduced in the beginning of Section 3, in our approach tracing links are contained into special models (technically called weaving models). This allows us (i) to keep the ADD model, architecture descriptions and requirements clearly separated and (ii) to be independent from any architectural language or requirements specification language. A weaving model can be seen as a means for setting fine-grained relationships between models and executing operations on them. The types of relations that can be expressed in our weaving models are shown in Figure 3.

The `TracingModel` metaclass represents the root of each design decision weaving model. It has a reference to the decision model, and a reference to the artifact to be linked (i.e., `decisionModel` and `artifact`, respectively) and contains a set of Links (*links* in Figure 3) which can be given in any order. Each `Link` represents a fine-grained traceability link between one or more ADDs and one or more element of the linked artifact; the "1..*" cardinality of these two relationships means that in this context

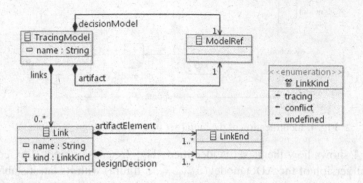

Fig. 3. Design decision weaving metamodel

traceability is a many-to-many relationship. The *kind* attribute specifies the type of the current link, it may be:

1. *tracing*: if the linked artifact elements trace to/from the linked ADDs;
2. *conflict*: if the linked artifact elements are not compliant with the linked ADDs;
3. *undefined*: if there is the need to link artifact elements and ADDs, but at the moment there is no way to establish whether it is a tracing or a conflict link.

It is important to note that by using our approach, designers are not forced to manually create weaving models, rather designers define traceability links by simply dragging and dropping the elements to be linked across the artifacts; then the corresponding weaving models are automatically created by the tool.

3.3 Evolution Impact Analysis

In Section 3 we explained how an ADD may evolve, and the basic steps of our approach in analysing the evolution impact. In this section we detail each step of the analysis.
Identification of the evolved ADD elements. The main goal of this step is to identify the evolved ADD elements that are relevant for the evolution impact analysis. The outcome of this step is E: the set of ADDs that must be checked by the designer, e.g., ADDs that have been either added, modified (e.g., status change), or that depend on other evolved elements (e.g., added issue or concern). In order to keep the approach well focussed on the changes that are relevant for evolution impact analysis, only some kinds of change within the ADD model are considered during analysis. For example, a change of the rationale of an ADD or a change of the description of an issue are not very much related to evolution impact analysis; the same holds for categories, stakeholders, and concerns. Referring to the $ADDMM$ metamodel, the kind of changes relevant for our evolution impact analysis are:

- Addition/deletion of an issue
- Addition/deletion of a concern
- Addition/modification/deletion of a design decision

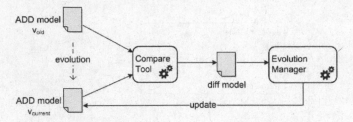

Fig. 4. Identification of the initial set of evolved design decisions

Figure 4 shows how the E set is identified. The E set is computed by comparing the current version of the ADD model ($v_{current}$ in figure) with its latest stable version (v_{old} in figure); the current prototype automatically keeps track of the last two versions of the ADD model, see Section 4.3. The comparison between the two version of the ADD model is performed by reusing an already existing model differencing tool, which produces as output a difference model (*diff model* in figure). Such a model enables timely identification of added, deleted or modified elements within the whole ADD model. The Evolution Manager component analyzes the *diff model*, filters out all the changes that are not relevant for our evolution impact analysis, and updates the current version of the ADD model by marking all design decisions in E as *evolved*. These design decisions will be the focus of next analysis steps.

Inter-decisions analysis. At this point, the approach establishes the ADDs in the model that depend on the ones in E, and adds them to E. The E set is populated with these extra design decisions as follows:

$$computeImpact(E) = E \cup \bigcup_{e \in E} computeImpact(dependsOn(e))$$

where $dependsOn(e)$ is the set of all design decisions depending on e. This dependency relationship is calculated as the union of all the other references of a specific design decision (e.g., *constraints*, *conflictsWith*, etc.).

The updated E set now represents all ADDs that must be checked with respect to the performed evolution. All the ADDs included into the E set are presented to the designer, so that she can focus and reason only on those impacted by the evolution. As stated at the beginning of Section 3, it is important to *validate* the current version of the ADD model; to this purpose, we defined a set of constraints that are used to semantically check the ADD model. These constraints are defined in the Object Constraint Language (OCL), an OMG standard language used to describe expressions on models. Listing 1.1 shows an example of such constraints: it considers each design decision in the ADD model (line 1 in the listing), and checks if the design decision (either directly or indirectly) conflicts with itself (line 3 in the listing).

```
1   context  DesignDecision
2       inv  conflictsWithSelf :
3           not  self.allConflicts()->exists(e | e = self)
```

Listing 1.1. Example of OCL contraint validating Design Decisions

Our approach defines different constraints like this, for example there is a constraint checking if there is some issue which is not addresses by any decision, there are constraints checking decisions with inconsistent statuses (e.g., a rejected decision which constraints an accepted one), and so on. Due to space limitations, we do not describe those constraints in this paper.

At this point, the ADD model will be corrected by the designer, and checked again in an iterative fashion, until the ADD model reaches a stable correct status. So, at the end of this step the ADD model is correct and its design decisions are consistent with each other.

Extra-decisions analysis. The scope of this part of the analysis is the whole set of involved artifacts (i.e., requirements and architectural descriptions). The main goal of this analysis is to identify which elements in other artifacts are impacted by the evolved design decision model. So, the weaving links connected to at least a design decision in the E set are shown to the designer in a different way. It is important to note that this analysis step is skipped if all design decisions in E are internal decisions (see metamodel in Section 3.1) since in this case there is no evolved design decision that is related to any external artifact.

Validation is an important issue also in this step; however, unlike inter-decisions validation, extra-decisions validation must consider the both the ADD model and other different artifacts (and their relationships) at the same time. This is achieved by specifying OCL constraints on weaving models, rather than on the ADD model itself. For example, Listing 1.2 shows an excerpt of OCL constraint that considers each *Link* contained into a weaving model between an SA description and a DD model (line 1), and checks if it is related to a *rejected* design decision (line 4 in the listing); this situation may be risky because the state of the decision may have changed to *rejected*, but the architectural element linked to that decision has not been updated accordingly.

```
1  context  Link
2      inv  rejectedDecisions  :  if(self.kind = LinkKind::tracing)
3           then
4                self.designDecision->forAll(e | e.state != '
                   rejected')
5           else  true
6           endif
```

Listing 1.2. Example of OCL constraint validating related artifacts

The result of this validation gives designers insights and accurate information about which element (being it an architectural element, or a requirement) is not aligned with the evolved design decisions. For example, there may be:

- an architectural component that is in conflict with a chosen decision,
- a requirement which is traced to a rejected decision,
- an architectural element which is traced to two mutually exclusive design decisions.

Such validation helps the designers to ensure the normal functioning of the system even during impairments caused by changing decisions because any conflicting decision is immediately flagged. It is important to note that, unlike many other approaches for design decision analysis, our approach allows to trace the evolution not only to the various impacted artifacts (for example, to a specific architecture description), but also to fine-grained elements of the artifacts (like specific architectural components, interfaces, and so on). It is up to the designer now to analyze the information provided by the analysis step and to correct the involved artifacts according to the evolved design decisions. Clearly, a change in one of the artifacts may imply a change back to the ADD model, triggering again a new design decisions evolution: this results in an iterative process in which design decisions evolution is iteratively evaluated and reflected to the involved artifacts. This process is part of the typical architectural reasoning activity.

4   Prototype Implementation

The current version of the prototype realizing the approach has been implemented as an Eclipse[1] plugin that relies on the Atlas Model Management Architecture (AMMA) [15]. We chose AMMA since it best fits our technical needs like (ADL and requirements) metamodel independence, high flexibility of the traceability metamodel, integration to Eclipse and its modeling facilities [16]. The next sections will detail the technologies we used in each part of the proposed approach.

4.1   Models and Metamodels Implementation

Models and metamodels are expressed via Ecore, the metamodeling language of the Eclipse Modeling Framework (EMF[2]), and serialized into standard XMI files. The technology we use for creating weaving models is the Atlas Model Weaver (AMW) [17]. In order to have a homogeneous framework we make the assumption that all heterogeneous artifacts that we use are models and that each model conforms to its metamodel. This enables the management of the tracing and validation of the involved artifacts since their elements are fully defined and encoded in the models. The assumption also seems reasonable in light of the recent Doc2Model (Document to Model[3]) Eclipse project for parsing structured documents to produce EMF models. We see this assumption as reasonable considering the benefits it offers.

4.2   Seamless Visualization of Architectural Artifacts

As previously stated, it is fundamental to clearly visualize the ADD model together with its related artifacts. So, while designing the tool we had two orthogonal concerns: (i) to show the ADD model together with other models in a seamless way, and (ii) to provide a means to easily define tracing links between all those models. In this work we propose a solution in which ADD models and their related artifacts can be displayed and edited simultaneously on different 3D planes and 3D links can be used to display inter-model connections. Figure 5 gives an overview on how the current version of the prototype renders ADDs and the other artifacts to designers. More specifically, it shows

---

[1] http://eclipse.org

[2] http://www.eclipse.org/modeling/emf

[3] http://eclipse.org/proposals/doc2model

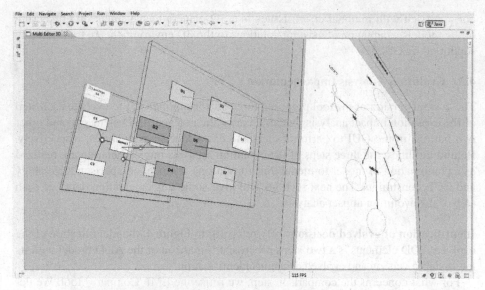

Fig. 5. Visualization of ADDs and their relationships to other artifacts

ADDs by means of the graphical notation we defined for the $ADDMM$ metamodel, requirements as a UML use case diagram and the architecture of the system as a UML component diagram. It is important to note that our approach is not dependent on UML; UML has been used only for illustrative purposes.

Requirements, ADDs and architecture descriptions are rendered on three separated planes in a 3D environment. ADDs are in the middle plane and in general are shown as white boxes. In Figure 5 there are some coloured decisions indicating the results of the evolution analysis: red ADDs are the evolved ones, and yellow ADDs are those depending on the evolved ones. The prototype automatically updates the ADD model by coloring the ADDs according to the results of the evolution analysis. The other planes are dedicated to requirements and architecture descriptions. It is important to note that designers may add other artifacts via additional planes; this can be done by simply dragging and dropping a (suitably adapted) diagram into the 3D environment.

Tracing links are visualized as connections between the various elements of the models displayed on these planes. Also, the colours of the connections play an important role: red connections indicate violations of some rule of the extra-decision analysis presented in Section 3.3, whereas the tracing links that passed the analysis are rendered as green.

From a technological point of view, the 3D environment and the visualization of the involved models on planes rely on GEF3D[4], an Eclipse modeling environment supporting 3D diagrams editing. GEF3D allows designers to create 3D diagrams, 2D diagrams and combine 3D with 2D diagrams (this is what our prototype currently does). An important feature of GEF3D is that existing Eclipse-based 2D editors can be easily embedded into 3D editors; so designers can basically reuse their preferred notations in

---

[4] http://www.eclipse.org/gef3d/

our approach with minimal effort. Issues like 3D camera management, 3D rendering of the models, adaptation of existing notations to the 3D environment are transparently supported by GEF3D.

4.3 Evolution Analysis Implementation

The current version of the tool proposed three buttons in the GUI to execute each step of the evolution impact analysis: one to identify the evolved ADD elements, and other two buttons in the GUI to perform inter- and extra- decisions analyses, respectively. Behind the lines, the three steps of the evolution analysis implementation are realized as a combination of model-to-model transformations, a model comparison technology and OCL constraints. The next sections will give some implementation detail of each step of the evolution impact analysis.

Identification of evolved decisions. By referring to Figure 4, the identification of the evolved ADD elements is a two-steps process: comparison of the ADD model with its latest stable version and evolved decisions identification.

For what concerns the comparison step, we reuse the EMF Compare[5] tool. We decided to use EMF Compare because it is well integrated in Eclipse, and provides a set of APIs to interact with its facilities. Among other features, EMF Compare includes a generic comparison engine and creates the corresponding delta; this delta may itself be represented as a model. We use this model as a means to represent differences between the current ADD model and its latest stable version (it is called *diff model* in Figure 4). Clearly, in our approach the intermediate *diff model* is kept in memory, and discarded after the results of the analysis are shown in the initial ADD model.

Evolved design decisions represent the E set described in Section 3.3 and are identified in the model by their `evolved` attribute set to *true*. The *Evolution Manager* component in Figure 4 is implemented as a model-to-model transformation that takes as input the ADD model and the *diff model*, selects design decisions are related to relevant evolved ADD elements, and updates the ADD model by setting to *true* the "evolved" attribute of all evolved decisions. This transformation is specified using the Atlas Transformation Language (ATL) [17], an hybrid model transformation language with declarative and imperative constructs.

Inter- and extra- decisions analyses implementation. In the current version of the prototype, inter- and extra- decisions analyses are executed by pressing two dedicated buttons in the GUI, one for each kind of analysis; this allows designers to easily check the current state of design decisions without interrupting their reasoning process. The results of the validation will be shown in the standard "Error Log" view of the Eclipse framework (see Figure 6 for a screenshot of the prototype showing a list of issues).

The identification of the element impacted by a design decision evolution is performed by means of a set of transformations similar to the one realizing the *Evolution Manager* component. The only difference is that in inter-decisions analysis, the E set is expanded with design decisions dependent on some decisions already in E, whereas in extra-decisions analysis, the transformation marks weaving links

---

[5] http://www.eclipse.org/emf/compare/

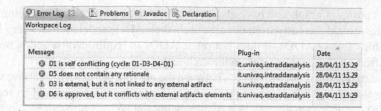

Fig. 6. The results of inter- and extra- decisions analyses

For what concerns the validation parts of the analyses, they are implemented as a set of OCL constraints specified using the Topcased-VF OCL evaluator[6]. Moreover, tracing links from/to design decisions to other related artifacts are rendered as red or grey, depending on the results of the validation; i.e., if some tracing link does not pass the inter-decisions analysis, it is rendered as red in order to helps designers to have a quick idea of the problems affecting current design artifacts. (see Figure 5 for a screenshot of the prototype showing this feature).

5   Related Work

The evolution of software systems was recognized as early as 1969 [18]. Since then, lot of efforts have been directed towards managing evolution in a effective manner. ISO/IEC 42010 standard on Architecture Description [14], provides direct correspondence between ADDs and Architecture Description elements which we have represented as tracing links.

Jansen and Bosch have proposed in [1] a tool that provides an architecture description language, a notation for representing design decisions, and a mechanism for integrating them. Similarly to our approach, their main goal is to keep ADDs as first-class entities and to strongly connect them to the architecture. However, they focus less on the various relationships that may exist either among ADDs or between ADDs and other artifacts. We believe that this plays a fundamental role in characterizing the evolution. Also, their approach depends on the Archium ADL itself, whereas our approach is ADL-independent.

Capilla et al. have proposed a metamodel for architecting and evolving architectural knowledge in [11]. The metamodel comprises three parts (the project model, the decision model and the architecture model) and they provide an *integrated view* combining requirements, ADDs and software architecture into a single process. Even if we share with Capilla et al. the same idea of an integrated view including requirements, ADDs and architecture, we focus more on characterizing how design decisions may evolve and on what is the impact of such an evolution either on other ADDs or other related artifacts. Furthermore, our approach allows designers to simultaneously visualize and better reason about ADDs and their relationships to other artifacts.

Authors of [5] propose a metamodel for ADDs and establish a 1:1, 1:n and n:1 relationship between ADDs and the elements. To handle evolution, they have used a

---

[6] http://gforge.enseeiht.fr/projects/topcased-vf

decision constraint graph and have also provide a change impact analysis using an example. However, differently from our approach, their ADD model includes very limited relationships among ADDs, it does not relate ADDs to requirements, and is not independent from the language to design the architecture.

Summarizing, the main differences between the approaches described above and our work are that (i) we use a generic and notation-independent metamodel to represent ADDs, (ii) our approach ensures that requirements and architecture can be represented using any preferred language, and (iii) our approach allows designers to simultaneously visualize and reason about ADDs and their relationships to other artifacts.

6   Conclusion and Future Work

Design decisions must be considered first-class elements when architecting software systems, and they have to be explicitly documented and supported. The fact that software architecture is a continuously evolving entity is not new, so explicit support and managing of the evolution of design decisions is becoming a necessity. The strong relationship that design decisions share with requirements and other artifacts needs to be taken into consideration while supporting evolution.

In this work we propose a model-driven approach for supporting the evolution of design decisions. Within the approach, our main contributions are (i) a metamodel for representing evolving design decisions; (ii) a means to create bidirectional traceability links between ADDs, requirements and artifacts which will help in analysing the impact of evolution on the artifacts; (iii) a technique to identify evolved ADDs and to check the impact of such an evolution both on other decisions and on other related artifacts.

As future work, we are investigating to make the proposed approach totally independent from the metamodel used to represent ADDs. This can be done by allowing designers to create models representing only the evolution of ADDs and then providing a mechanism to bind those models on the various models used to represent ADDs. Furthermore, it would be interesting to enhance the evolution analysis part with some mechanism to (semi)automatically solve the identified problems (e.g., by applying some resolution pattern to each constraint running during the analysis).

An interesting and important increment to our work will be to add quality aspects to the model. Though currently we take into account stakeholder concerns, we have not explicitly accounted for non-functional aspects like scalability, reliability, dependendabilty etc in out model. By including these, we believe that our model will be better capable of handling evolution thereby ensuring that the system retains functionality even while undergoing changes.

Finally, the completeness of the tracing relationship types, and of the set of constraints executed during the evolution analysis will be experimented on a concrete, industrial case study. This will also help in understanding possible usability and performance issues related to the tool implementing the approach.

References

1. Jansen, A., Bosch, J.: Software architecture as a set of architectural design decisions. In: WICSA 2005 (2005)

2. Kruchten, P.: An Ontology of Architectural Design Decisions in Software Intensive Systems. In: 2nd Groningen Workshop Software Variability, pp. 54–61 (2004)
3. Potts, C., Bruns, G.: Recording the reasons for design decisions. In: 10th International Conference on Software Engineering ICSE 1988, pp. 418–427 (1988)
4. Capilla, R., Nava, F., Tang, A.: Attributes for characterizing the evolution of architectural design decisions. In: 2007 Third International IEEE Workshop on Software Evolvability, pp. 15–22 (2007)
5. Choi, Y., Choi, H., Oh, M.: An architectural design decision-centric approach to architectural evolution. In: 11th International Conference on Advanced Communication Technology, ICACT 2009, vol. 01, pp. 417–422 (2009)
6. Gilson, F., Englebert, V.: An architectural design decision-centric approach to architectural evolution. In: 5th Workshop on Variability Modeling of Software-Intensive Systems (2011)
7. Mens, T., Magee, J., Rumpe, B.: Evolving software architecture descriptions of critical systems. Computer 43, 42–48 (2010)
8. Hofmeister, C., Kruchten, P., Nord, R.L., Obbink, H., Ran, A., America, P.: A general model of software architecture design derived from five industrial approaches. Journal of Systems and Software 80(1), 106–126 (2007)
9. Tang, A., Liang, P., Clerc, V., van Vliet, H.: Supporting co-evolving architectural requirements and design through traceability and reasoning. In: Relating Software Requirements and Architectures. Springer, Heidelberg (2011)
10. Capilla, R., Nava, F., Perez, S., Duenas, J.C.: A web-based tool for managing architectural design decisions. ACM SIGSOFT Software Engineering Notes 31(5) (2006)
11. Capilla, R., Nava, F., Duenas, J.C.: Modeling and documenting the evolution of architectural design decisions. In: Second Workshop on SHAring and Reusing Architectural Knowledge Architecture, p. 9 (2007)
12. Levendovszky, T., Rumpe, B., Schätz, B., Sprinkle, J.: Model evolution and management. In: Giese, H., Karsai, G., Lee, E., Rumpe, B., Schätz, B. (eds.) Model-Based Engineering of Embedded Real-Time Systems. LNCS, vol. 6100, pp. 241–270. Springer, Heidelberg (2010)
13. Eklund, U., Arts, T.: A classification of value for software architecture decisions. In: Babar, M.A., Gorton, I. (eds.) ECSA 2010. LNCS, vol. 6285, pp. 368–375. Springer, Heidelberg (2010)
14. ISO: Final committee draft of Systems and Software Engineering – Architectural Description (ISO/IECFCD 42010). Working doc.: ISO/IEC JTC 1/SC 7 N 000, IEEE (2009)
15. Bézivin, J., Jouault, F., Rosenthal, P., Valduriez, P.: Modeling in the large and modeling in the small. In: Aßmann, U., Aksit, M., Rensink, A. (eds.) MDAFA 2003. LNCS, vol. 3599, pp. 33–46. Springer, Heidelberg (2005)
16. Bézivin, J., Bouzitouna, S., Del Fabro, M., Gervais, M.P., Jouault, F., Kolovos, D., Kurtev, I., Paige, R.F.: A canonical scheme for model composition (2006)
17. Jouault, F., Kurtev, I.: Transforming Models with ATL. In: Proceedings of the MTP, Workshop at MoDELS 2005 (2006)
18. Lehman, M.: The programming process. IBM Research Report (1969)

On Enabling Dependability Assurance in Heterogeneous Networks through Automated Model-Based Analysis

Paolo Masci[1,*], Nicola Nostro[2], and Felicita Di Giandomenico[2]

[1] Queen Mary University of London, UK
paolo.masci@eecs.qmul.ac.uk
[2] ISTI-CNR, Pisa, Italy
{nicola.nostro,felicita.digiandomenico}@isti.cnr.it

Abstract. We present the specification of a basic library of dependability mechanisms that can be used within automated approaches for synthesising dependable CONNECTors in heterogeneous networks. The library builds on classical dependability patterns, such as majority voting and retry, and uses the concept of overlay networks for triggering the synthesis of specific dependability mechanisms in the CONNECTor from high-level specifications. We translated such dependability mechanisms into SAN models with the aim to evaluate, through model-based analysis, which dependability mechanisms should be embedded in the synthesised CONNECTor for ensuring a given dependability level between networked systems willing to be connected. A case study is also presented to show the application of selected library mechanisms. This work is carried out in the context of CONNECT, a European FET project which is investigating the possibility of enabling long-lasting inter-operation among networked systems by synthesising mediating CONNECTors at run-time.

1 Introduction and Background

Interoperability in future networks will be characterised by seamless and continuous communication among heterogeneous networked systems. Over time, networked systems may change mode of operation, e.g., because of hardware/software updates or new application contexts. As a consequence, network infrastructures ought to provide appropriate means for supporting interoperability among evolving networked systems.

In the European FET project CONNECT [1], interoperability issues of evolving networks are tackled by synthesising dependable CONNECTors at run-time. To do this, the network infrastructure defined in CONNECT embeds five logical units, denominated *enablers*, that seamlessly collaborate for ensuring continuous and long-lasting inter-operation among networked systems. In CONNECT, the *discovery enabler* discovers the functionality of networked systems and applications and retrieves information on the interfaces they use for inter-operating with others.

---

* Corresponding author.

E.A. Troubitsyna (Ed.): SERENE 2011, LNCS 6968, pp. 78–92, 2011.

The *learning enabler* completes such a knowledge on the interaction behaviour of networked systems by applying learning algorithms, and produces a model of this behaviour in the form of a labelled transition system (LTS). The *synthesis enabler* dynamically synthesises a software mediator using code generation techniques (from the independent LTS models of each system) that will connect and coordinate the interoperability between heterogeneous systems. In order to fulfil dependability requirements, *synthesis* triggers the *dependability enabler*, which is in charge of analysing the CONNECTor's design before the CONNECTor gets deployed and put in operation. If needed, the dependability enabler drives the synthesis enabler towards possible CONNECTor's enhancement. The *monitoring enabler* continuously monitors the deployed CONNECTors during their execution for updating the other enablers with run-time data.

In this work, we focus on the dependability enabler, which performs a model-based analysis for assessing the dependability level of the synthesised CONNECTors. Specifically, we point our attention on a dependability enabler's module, denominated *enhancer*. Such a module is responsible for guiding the synthesis process towards enhancements of a CONNECTor's design whenever the analysis reveals inadequate dependability levels. In brief, this module is in charge of selecting a combination of dependability mechanisms suitable for enhancing the synthesised CONNECTor so that it complies with given requirements. Then, the *synthesis enabler* embeds the selected dependability mechanisms in the CONNECTor's design and proceeds with its implementation and deployment. The architecture of the dependability enabler and the functionalities of the enhancer have been presented in [13]. The contribution of this paper consists in the definition of a basic library of dependability mechanisms for the enhancer. The library builds on classical dependability patterns (see [18] for a survey), and uses the concept of *overlay networks* for triggering the synthesis of specific dependability mechanisms in the synthesised CONNECTor from high-level specifications. How these dependability mechanisms can then be embedded in the synthesised connectors pertains to the synthesis enabler and is not addressed in this paper.

The paper is organised as follows. In Section 2, the stochastic activity networks [15] (SAN) formalism, a widely used formalism for model-based dependability analysis of complex systems, is briefly illustrated in order to allow the reader to understand the formal specification of the developed dependability mechanisms. In Section 3, we explain the ideas underpinning the library of dependability mechanisms, and we present the specification of such a library with the SAN formalism. In Section 4, we trial our ideas by applying the library to a case study based on a demonstrative scenario based on that presented in [8]. In Section 5, we report on related work and conclude the paper.

2   The SAN Formalism

Stochastic Activity Networks [15,14,21] are an extension of the Petri Nets (PN) formalism [17,16]. SANs are directed graphs with four disjoint sets of nodes: *places*, *input gates*, *output gates*, and *activities*.

Activities replace and extend the *transitions* of the PN formalism. Each SAN activity may be either *instantaneous* or *timed*. Timed activities represent actions with a duration affecting the performance of the modelled system, e.g., message transmission time. The duration of each timed activity is expressed via a *time distribution* function. Any instantaneous or timed activity may have mutually exclusive outcomes, called *cases*, chosen probabilistically according to the *case distribution* of the activity. Cases can be used to model probabilistic behaviours. An activity *completes* when its (possibly instantaneous) execution terminates.

As in PNs, the state of a SAN is defined by its *marking*, i.e., a function that, at each step of the net's evolution, maps the places to non-negative integers (called the *number of tokens* of the place). SANs enable the user to specify any desired enabling condition and firing rule for each activity. This is accomplished by associating an *enabling predicate* and an *input function* to each input gate, and an *output function* to each output gate. The enabling predicate is a Boolean function of the marking of the gate's input places. The input and output functions compute the next marking of the input and output places. If these predicates and functions are not specified for some activity, the standard PN rules are assumed. The evolution of a SAN, starting from a given marking μ, may be described as follows:

1. the instantaneous activities enabled in μ complete in some unspecified order;
2. if no instantaneous activities are enabled in μ, the enabled (timed) activities become *active*;
3. the completion times of each active (timed) activity are computed stochastically, according to the respective time distributions; the activity with the earliest completion time is selected for completion;
4. when an activity (timed or not) completes, one of its cases is selected according to the case distribution, and the next marking μ' is computed by evaluating the input and output functions;
5. if an activity that was active in μ is no longer enabled in μ', it is removed from the set of active activities.

Graphically, places are drawn as circles, input gates as left-pointing triangles, output gates as right-pointing triangles, instantaneous activities as narrow vertical bars, and timed activities as thick vertical bars. Cases are drawn as small circles on the right side of activities.

3   Library of Dependability Mechanisms

Overlay networks are virtual networks built on top of existing network substrates: nodes in overlay networks represent logical hosts involved in interactions, and links in overlay networks correspond to paths in the network substrate that are traversed by messages during inter-operations. To date, overlay networks have been used for exploiting peculiar characteristics of network substrates at the application level; for instance, Andersen et al [3] exploit highly redundant networks substrates for defining resilient communication systems as overlay networks (a survey on the use of overlay networks for defining new applications can be found in [12]).

Thanks to the infrastructure provided by CONNECT, here we can use the concept of overlay networks in an alternative way, i.e., rather than using overlay networks for exploiting the characteristics of the network substrate, we use them for *defining* the characteristics of the network substrate that should be synthesised. The basic idea is to view CONNECTors as overlay networks, and to exploit their structure for triggering the generation of specific dependability mechanisms in the network substrate during the synthesis process.

In the following, we describe the models we defined for triggering the generation of typical dependability mechanisms suitable to contrast two typical classes of failure modes that may happen during interactions: *timing failures*, in which networked systems send messages at time instants that do not match an agreed schedule, and *value failures*, in which networked systems send messages containing incorrect information items. For the purpose of this paper, we consider timing failures of type omission, i.e., late messages are always discarded, and value failures that cause a networked system to respond within the correct time interval but with an incorrect value. Each model is specified with the SAN formalism. The models are developed according to three basic rules that allow to simplify the automated procedure for embedding the mechanism in the specification of the synthesised CONNECTor: (i) each model has an initial place, s0, whose tokens enable the first activity of the model; (ii) each model has a final place, s1, which contains tokens whenever the last activity of the model completes; (iii) the overall number of tokens in s1 is always less or equal to the number of tokens in s0. With the above rules, the behaviour of the model can be seen as an *enhanced activity*, and can be directly used to replace any activity that moves tokens between two places in the specification of the CONNECTor (the basic semantics of an activity is always preserved).

3.1   Retry Mechanism

The retry mechanism consists in re-sending messages that get corrupted or lost during communications, e.g., due to transient failures of communication links. This mechanism is widely adopted in communication protocols, such as TCP/IP [6] for enabling reliable communication over unreliable channels. A typical implementation of the retry mechanism uses time-outs and acknowledgements: after transmitting a message, the sender waits for a message of the receiver that acknowledges successful communication. If the acknowledgement is not received within a certain time interval, the sender assumes that the communication was not successful, and re-transmits the message.

The synthesis of a retry mechanism can be triggered with the stochastic activity network shown in Figure 1. On the sender side, the mechanism creates a message re-transmission policy for re-sending the message at most N times; on the receiver side, the mechanism creates a policy for avoiding duplicated reception of messages and for sending acknowledgements.

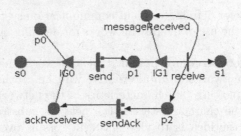

Fig. 1. Retry mechanism

In the model, all places initially contain zero tokens, except p0, which contains
N tokens, where N is a model parameter representing the maximum number of
re-transmissions. Activity send is enabled when the conjunction of the following conditions is true: p0 and s0 contain at least one token, and ackReceived
contains zero tokens. When activity send completes with success (case 0, with
probability $pr0$), a token is removed from s0 and p0, and the marking of p1
is incremented by one. Activity receive is enabled when p1 contains at least
one token and messageReceived contains zero tokens. When activity receive
completes, a token is moved to s1, and the marking of p2 and messageReceived
is incremented by one. A token in p2 enables activity sendAck, whose aim is to
enable the receiving host notify the sender that the message has been successfully received. The sender stops re-transmitting the message as soon as it gets
an acknowledgement that the message has been successfully received, or after N
attempts.

3.2   Probing Mechanism

The probing mechanism exploits redundant paths and periodic *keep-alive* messages for enabling reliable communication in face of path failures. The basic idea
is to continuously collect statistics on the characteristics of the communication
channels, and to select the best channel on the basis of such statistics. This
mechanism has been used for defining communication services with guaranteed
delivery and performance levels, e.g., see Akamai's SureRoute [2] and reliable
multi-cast protocols for peer-to-peer networks [23].

The synthesis of a probing mechanism that uses two redundant communication channels can be triggered with the stochastic activity network shown in
Figure 2. The mechanism instruments the sender with a periodic channel probing
functionality suitable to feed a monitoring system that collects statistics about
the reliability level of the communication channels.

In the model, place mode is a state variable that indicates the mode of operation of the mechanism, which can be either *probing mode* (mode contains zero
tokens), i.e., the mechanism tests the characteristics of the communication channels through keep-alive messages, or *normal mode* (mode contains one token), i.e.,
the mechanism selects the best estimated channel for relaying messages. Initially,
all places contain zero tokens, except ready, which contains one token.

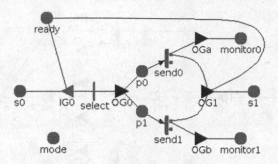

Fig. 2. Probing mechanism

When in normal mode, activity `select` is enabled when `s0` contains at least one token, and `ready` contains one token. When `select` completes, one token is removed from `s0` and `ready`, and `send0` gets enabled if `monitor0` has more tokens than `monitor1` (`send1` gets enabled in the other case). If `send0` completes with success (case 0), then a token is added to `s1` and `ready`. Similarly, when `send1` completes with success, a token is moved to `s1` and `ready`.

When in probing mode, the model behave as follows: `send0` and `send1` have both the same rate $R0$ (while their case probabilities depend on the characteristics of the channels, which may vary over time). Activity `select` is enabled when `ready` contains one token; when `select` completes, `ready` contains zero tokens and activities `send0` and `send1` get enabled (by moving one token in both `p0` and `p1`). When `send0` completes with success (case 0), a token is added to `monitor0`. Similarly, when `send1` completes, a token is added to `monitor1`. A token is moved to `ready` when both `send0` and `send1` complete.

3.3   Majority Voting Mechanism

Majority voting is a fault-tolerant mechanism that relies on a decentralised voting system for checking the consistency of data. Voters are software systems that constantly check each other's results, and has been widely used for developing resilient systems in the presence of faulty components. In a network, voting systems can be used to compare message replicas transmitted over different channels, see, for instance, the protocol proposed in [24] for time-critical applications in acoustic sensor networks.

The synthesis of a majority voting mechanism that uses three redundant communication channels can be triggered with the stochastic activity network shown in Figure 3. The mechanism replicates the message sent by the transmitting host over three channels. In this case, the mechanism is able to tolerate one faulty channel.

In the model, all places initially contain zero tokens. Activity `multipathRouter` gets enabled when `s0` contains a token. When such an activity completes, the token is removed from `s0`, and three send activities (`send0`, `send1`, `send2`) get enabled by moving tokens into places `p0`, `p1`, and `p2`. The number of tokens moved in such places encode the actual informative content of the message.

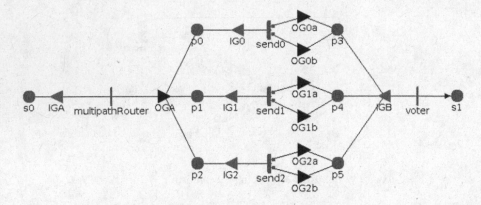

Fig. 3. Majority voting mechanism

When a send activity completes with success, the activity preserves the number of tokens (i.e., all tokens are moved forward into the next place). Activity **voter** gets enabled when the all sends complete (such activities will eventually change the marking of **p3**, **p4**, and **p5**). When activity **voter** completes, a token is moved into **s1** and all tokens in other places are removed.

3.4 Error Correction Mechanism

Error correction deals with the detection of errors and re-construction of the original, error-free data. A widely used approach for enabling hosts to automatically detect and correct errors in received messages is *forward error correction* (FEC). The mechanism requires the sender host to transmit a small amount of additional data along with the message. The mechanism has been used, for instance, in [22] for defining an overlay-based architecture for enhancing the quality of service of best-effort services over the Internet.

The synthesis of an error correction mechanism that uses two redundant communication channels can be triggered with the stochastic activity network shown in Figure 4. One channel is used to send the original message, and the other channel is used to send the error correction (EC) code. The receiver is instrumented with a filtering mechanism that checks and corrects messages.

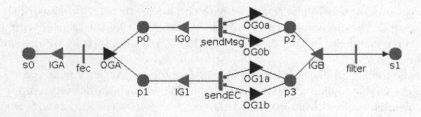

Fig. 4. Error correction mechanism

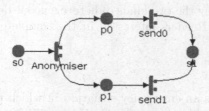

Fig. 5. Security mechanism

Initially, all places contain zero tokens. When a token is moved into s0, activity fec gets enabled. When such an activity completes, a token is removed from s0, and activities sendMsg and sendEC get enabled by moving tokens into p0 and p1. The number of tokens moved in such places encode the actual informative content of the message. When sendMsg completes with success, all tokens in place p0 are moved into p2. Similarly, when sendEC completes, all tokens of p1 are moved into p3. Activity filter gets enabled when places p2 and p3 contain tokens. When activity filter completes, a token is moved into s1 and all tokens in other places are removed.

3.5 Security Mechanism

A typical way to enforce protection on a host is to decouple the host from the rest of the network. A host can, for instance, be protected from receiving unwanted traffic by creating a ring that selectively filters the incoming traffic. Similarly, the identity of a host can be protected by anonymising the host's messages through a set of intermediary hosts, denominated *proxies*. This mechanism has been used, for instance, in [11] and [4] for protecting hosts from denial-of-service attacks.

The synthesis of a security mechanism over a network with two intermediary hosts can be triggered with the stochastic activity network shown in Figure 5. The mechanism creates an anonymiser service that selects a channel with a certain probability, and forwards the message on such a channel. In the model, all places initially contain zero tokens. When a token is moved into s0, activity Anonymiser gets enabled. When such an activity completes, a token is moved from s0 either to p0 (with probability $pr0$), or to p1 (with probability $pr1$), and either send0 or send1 gets enabled. When a send activity completes, a token is moved in s1.

4   Case Study

In this section, we show how the developed library can be used within an automated model-based dependability analysis with the aim to enhance a synthesised CONNECTor. We consider a demonstrative scenario described in [8], where two kinds of heterogeneous devices need to communicate in a reliable and timely manner. For clarity of exposition, in order to fit the purpose of this paper and also make it self-contained, in the following we report a concise and slightly reworded description of the scenario reported in [8], and an informal specification of the protocols used by the two kinds of heterogeneous devices, and of the synthesised CONNECTor. Readers interested in the complete original specification are re-directed to [8].

4.1 Specification

The scenario considers an emergency situation in which policemen need to exchange information with security guards. Each policeman can exchange confidential data with other policemen with a *secured file sharing* protocol. Security guards, on the other hand, exchange information by using another protocol, denominated *emergency call*. The two protocols have the same aim (i.e., enable information exchange) but a mediating CONNECTor is needed in order to enable inter-operation, because they use different message types and different message sequences.

Secured File Sharing. This is a basic peer-to-peer protocol for enabling file sharing. The peer that initiates the communication, denominated *coordinator*, sends a multicast message (`selectArea`) to a selected group of peers. When a peer receives a `selectArea` message, the peer replies to the coordinator with a `areaSelected` message. Upon receiving the `areaSelected` message, the coordinator sends a data file to the peers (`uploadData` message) that, in turn, automatically reply with an `uploadSuccess` message if the data has been successfully received.

Emergency Call. This is a peer-to-peer protocol for sending data files from a control centre to groups of devices. Each group of devices is coordinated by a leader. The protocols is initiated by the control centre, which sends an `eReq` message to a group of devices located in a selected area of interest. The group leader is in charge of replying to the control centre with an `eResp`. Whenever the control centre receives the `eResp` from a group leader, an `emergencyAlert` message is sent to all devices. Each device automatically notifies the control center with an `eACK` message whenever it successfully receives the data.

Mediating Connector. A mediating CONNECTor suitable for enabling inter-operation from devices using the secured file sharing protocol to devices using the emergency call protocol performs the following translations: `selectArea` messages are translated into `eReq` messages directed to the leaders of selected group of devices; `eResp` messages are translated into `areaSelected` messages; `uploadData` messages are translated into multicast `emergencyAlert` messages; `eACK` messages are collected by the CONNECTor and then translated into a single `uploadSuccess` message. A timeout is used to avoid infinite wait in the case of failure of `eACK` messages.

4.2 SAN Models

The SAN models corresponding to the specifications are shown in Figure 6. The CONNECTor uses two different channels to communicate with the two different kinds of devices. Any send and receive action performed by the CONNECTor is

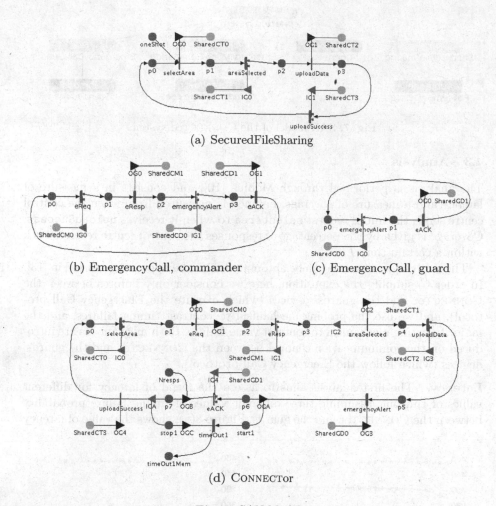

(a) SecuredFileSharing

(b) EmergencyCall, commander

(c) EmergencyCall, guard

(d) CONNECTor

Fig. 6. SAN Models

represented by a timed activity. Each activity has two case probabilities: case 0 is associated to the correct behaviour; case 1 is associated to incorrect behaviour. Since the purpose of this case study is not to show new results on a real-world protocol, but to exemplify the utility of the developed library, here we assume that timed activities are all exponentially distributed and with the same rate.

The model of the CONNECTed system, which is shown in Figure 7, is obtained by composing the models through place sharing. In the CONNECTed system, there is a shared place for each pair of activities that represent send/receive actions: send activities add tokens in the shared place, while receive activities remove tokens from the shared place and use the marking of the shared place as enabling condition.

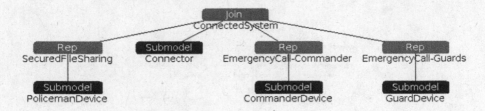

Fig. 7. SAN model of the CONNECTed system

4.3 Analysis

The analysis is performed through Möbius [10], and consists in a measure of latency and a measure of coverage. Latency is measured from when the control centre sends the initial request `selectArea` to when it receives `uploadSuccess`. Coverage is given by the percentage of responses the control centre receives back within a certain time T.

The analysis we describe can be automated with the approach reported in [13]. In order to simplify the exposition, here we consider only failures between the CONNECTor and the guards' devices (which execute the Emergency Call protocol), and we use the probing mechanism to contrast timing failures, and the majority voting mechanism to contrast value failure. Both mechanisms are introduced on the communication channel between the CONNECTor and the guards' devices (which follow the Emergency Call protocol).

Latency. The first analysis aims to assess the trend of latency for different values of timeout, assuming three different values of timing failure probability between the CONNECTor and the guards. Figure 8(a), shows the value of latency

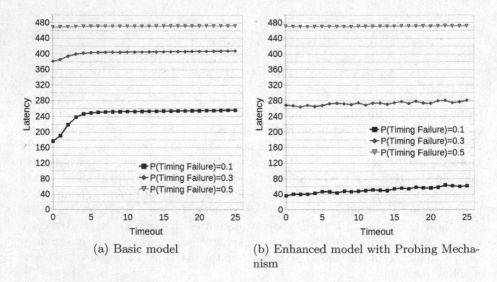

(a) Basic model

(b) Enhanced model with Probing Mechanism

Fig. 8. Latency assessment in case of timing failure

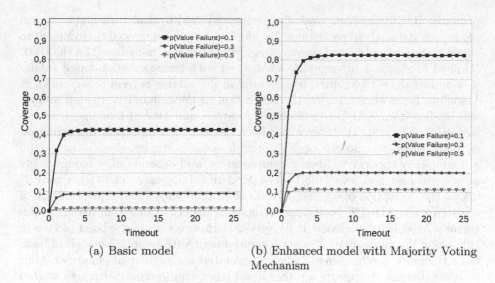

(a) Basic model   (b) Enhanced model with Majority Voting
                              Mechanism

Fig. 9. Coverage assessment in case of value failure

(on the y axis) for the CONNECTed system without dependability mechanisms
(the timeout value is reported on the x axis). Figure 8(b), shows the same analysis performed on the model enhanced with the probing mechanism. We can
notice that, with the considered system parameters, the mechanism is able to
reduce latency only in two out of three situations. When the timing failure probability is 0.5 the latency has very similar values for both models; in fact, in such
a case the failure probability has a too high value and the probing mechanism
is a too light means to contrast its effects.

Coverage. The second analysis is performed for three different probabilities of
value failures between the CONNECTor and the guards. Figures 9(a) and 9(b)
show the analysis results for the basic model and for the model enhanced with
the majority voting mechanism. In this case, the mechanism is able to improve
coverage in all considered cases. We can notice that, for the probability values considered, the coverage provided by the enhanced model is approximately
double compared to basic one.

5   Related Work and Conclusions

Automated dependability analysis, pursued through transformation-based verification and validation environments, has been the subject of several studies in the
last decade. Automatic/automated methods from system specification languages
to modelling languages amenable to perform dependability analysis has been
recognised as an important support for improving the quality of systems. Moreover, it favours the application of verification and validation techniques at industry level, where these methods have difficulties to be applied on a routine basis,

primarily due to the high level of expertise required to deal with mathematical modelling and analysis techniques. Development of an integrated environment to support the early phases of system design, where design tools based on the UML (Unified Modeling Language) are augmented with transformation-based validation and analysis techniques, is presented in [5], among several other works. A Modeling framework allowing the generation of dependability-oriented analytical models from AADL (Architecture Analysis and Design Language) models is described in [19]. Tools have also been developed, supporting the definition of model-based transformations. To provide some examples, the Viatra tool [9] automatically checks consistency, completeness, and dependability requirements of systems designed using the Unified Modeling Language. The Genet tool [7] allows the derivation of a general Petri net from a state-based representation of a system. The ADAPT Tool supports model transformations from AADL Architectural Models to Stochastic Petri Nets [20]. However, from the point of view of enhancing the model-transformation environment with template models of basic fault tolerance mechanisms to allow automated assessment of enhanced, fault tolerant designs, it appears a rather novel research direction. Although studies exist dealing with libraries of fault tolerance mechanisms (e.g. [18]) to assist the design of dependable systems, to the best of the authors' knowledge the attempt to provide template models of dependability mechanisms, to be incorporated in a wider system dependability model to assess their efficacy at system design time, is a new contribution of this paper.

In this paper, five dependability mechanisms have been specified in terms of SAN models, covering basic means to cope with timing and value failures of communication channels in heterogeneous networked systems. They are first encapsulated in the dependability model of the CONNECTor set-up to allow interoperability among networked systems, and managed by the dependability evaluator enabler to analyse their appropriateness to satisfy dependability properties required by the networked systems. Upon positive assessment, they are employed to build advanced CONNECTors design. A case study has been also included, inspired by current research activity ongoing in the context of the EU CONNECT project, to show the practical application of selected dependability mechanisms in presence of failure scenarios.

The work described is a first step in the development of enhanced automated dependability analysis as a support for the synthesis of dependable CONNECTors. After the definition of the individual dependability mechanisms, all the implications related with their systematic usage to replace basic elements of the CONNECTor dependability model (showing unsatisfactory from the dependability or performance point of view), have to be rigorously considered and solved. Of course, also investigations on further dependability mechanisms suitable to the addressed context would be interesting to carry-on. Indeed, these are among the directions we are exploring as future work.

Acknowledgements. This work is partially supported by the EU FP7 Project CONNECT (FP7–231167).

References

1. Connect: Emergent connectors for Eternal Software Intensive Networked Systems (2009-2013), http://connect-forever.eu/
2. Akamai Technologies, Inc. Akamai sureroute for failover and performance (2003)
3. Andersen, D., Balakrishnan, H., Kaashoek, F., Morris, R.: Resilient overlay networks. SIGOPS Oper. Syst. Rev. 35, 131–145 (2001)
4. Andersen, D.G.: Mayday: Distributed Filtering for Internet Services. In: 4th Usenix Symposium on Internet Technologies and Systems, Seattle, WA (March 2003)
5. Bondavalli, A., Cin, M.D., Latella, D., Majzik, I., Pataricza, A., Savoia, G.: Dependability analysis in the early phases of uml-based system design. Language 16(5), 265–275 (2001)
6. Braden, R.T.: RFC 1122: Requirements for Internet hosts—communication layers (October 1989)
7. Carmona, J., Cortadella, J., Kishinevsky, M.: Genet: A tool for the synthesis and mining of petri nets. In: ACSD 2009, pp. 181–185. IEEE Computer Society, Washington, DC, USA (2009)
8. CONNECT Consortium. Deliverable D5.2 – Dependability Assurance (available soon) (2011)
9. Csertan, G., Huszerl, G., Majzik, I., Pap, Z., Pataricza, A., Varro, D., Varró, D.: Viatra - visual automated transformations for formal verification and validation of uml models. In: 17th IEEE International Conference on Automated Software Engineering (ASE 2002), pp. 267–270 (2002)
10. Daly, D., Deavours, D.D., Doyle, J.M., Webster, P.G., Sanders, W.H.: Möbius: An extensible tool for performance and dependability modeling. In: Haverkort, B.R., Bohnenkamp, H.C., Smith, C.U. (eds.) TOOLS 2000. LNCS, vol. 1786, pp. 332–336. Springer, Heidelberg (2000)
11. Keromytis, A.D., Misra, V., Rubenstein, D.: Sos: Secure overlay services. In: Proceedings of ACM SIGCOMM, pp. 61–72 (2002)
12. Kurian, J., Sarac, K.: A survey on the design, applications, and enhancements of application-layer overlay networks. ACM Comput. Surv. 43, 5:1–5:34 (2010)
13. Masci, P., Martinucci, M., Di Giandomenico, F.: Towards automated dependability analysis of dynamically connected systems. In: Proc. 10th Intl. Symposium On Autonomous Decentralised Systems, ISADS 2011 (2011)
14. Movaghar, A.: Stochastic activity networks: a new definition and some properties. Scientia Iranica 8(4), 303–311 (2001)
15. Movaghar, A., Meyer, J.F.: Performability modelling with stochastic activity networks. In: Proc. of the 1984 Real-Time Systems Symposium, pp. 215–224 (1984)
16. Murata, T.: Petri nets: Properties, analysis and applications. Proceedings of the IEEE 77(4), 541–580 (1989)
17. Petri, C.A.: Kommunikation mit Automaten. PhD thesis, Technische Hochschule Darmstadt (1961)
18. ReSIST Consortium. EU project ReSIST: Resilience for Survivability in IST. Deliverable D33: Resilience-explicit computing. Technical report (2008), http://www.resist-noe.org/
19. Rugina, A.-E., Kanoun, K., Kaâniche, M.: A system dependability modeling framework using AADL and gSPNs. In: de Lemos, R., Gacek, C., Romanovsky, A. (eds.) Architecting Dependable Systems IV. LNCS, vol. 4615, pp. 14–38. Springer, Heidelberg (2007)

20. Rugina, A.-E., Kanoun, K., Kaaniche, M.: The adapt tool: From aadl architectural models to stochastic petri nets through model transformation. In: Seventh European Dependable Computing Conference, pp. 85–90 (2008)
21. Sanders, W.H., Meyer, J.F.: Stochastic activity networks: formal definitions and concepts. In: Lectures on Formal Methods and Performance Analysis: First EEF/Euro Summer School on Trends in Computer Scienc, pp. 315–343. Springer-Verlag New York, Inc., New York (2002)
22. Subramanian, L., Stoica, I., Balakrishnan, H., Katz, R.H.: Overqos: offering internet qos using overlays. SIGCOMM Comput. Commun. Rev. 33, 11–16 (2003)
23. Zeng, W., Zhu, Y., Lu, H., Zhuang, X.: Path-diversity p2p overlay retransmission for reliable ip-multicast. IEEE Transactions on Multimedia 11(5), 960–971 (2009)
24. Zhou, Z., Peng, Z., Cui, J.-H., Shi, Z.: Efficient multipath communication for time-critical applications in underwater acoustic sensor networks. IEEE/ACM Trans. Netw. 19, 28–41 (2011)

Supporting Cross-Language Exception Handling When Extending Applications with Embedded Languages

Anthony Savidis[1,2]

[1] Foundation for Research and Technology Hellas, Institute of Computer Science
[2] University of Crete, Department of Computer Science
as@ics.forth.gr

Abstract. Software applications increasingly deploy scripting embedded languages for extensibility, letting introduce custom extensions on top of the original application components. This results in two language layers, the one in which the application is implemented and the embedded language itself in which the extensions are written. During runtime, active calls amongst these two layers may be naturally intermixed, posing challenges for intuitive cross-language exception handling. At present, besides all .NET languages which cooperate by relying on a common language infrastructure, cross-language exception handling is not supported by existing embedded languages like Python, Perl, Ruby and Lua. We discuss the requirements for cross-exception handling and we show how they are accommodated via small-scale amendments in the embedded language API and runtime.

Keywords: Exception Handling, Cross-Language Exceptions, Embedded Languages, Application Extensions.

1 Introduction

The adoption of embedded languages, also commonly known as scripting languages, is a very popular method to facilitate post-production application extensions. It was firstly practiced at a large scale by game companies (Walker et al., 2009) and was quickly spread to many different application domains (Loui, 2008). Such extensions may be freely introduced by application users, while they are frequently produced by application vendors as valued application add-ons.

Such applications evolve by integrating extension components not originally designed and tested to interact with each other. The latter is generally acknowledged to raise fault-tolerance issues at an architectural level of abstraction where exception handling may play an important role (Brito et al., 2009). Today, the diversity of application development and delivery disciplines resulted in adapted, optimally fit, exception handling policies, targeted in addressing various aspects such as concurrency (Xu et al., 2000) and mobility (Iliasov & Romanovsky, 2005). Similarly, custom exception handling demands emerge for applications which are extensible by embedded languages. The primary reason is the presence of two language layers, the one in which the application is implemented and the embedded language itself in which the extensions are written. During runtime, active calls amongst these two

E.A. Troubitsyna (Ed.): SERENE 2011, LNCS 6968, pp. 93–99, 2011.

layers may be naturally intermixed, posing challenges for intuitive cross-language exception handling. The high-level extensibility architecture for applications is illustrated under Fig. 1. Applications may be extended using component technologies (*native extensions*) and embedded language technologies (*scripted extensions*). Regarding the former, exception handling for component-based systems can be effectively managed through special software engineering approaches (Romanovsky, 2001). We show that for the latter we need introduce extra exception handling features to be introduced within embedded languages.

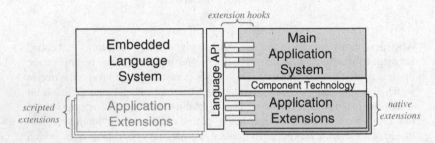

Fig. 1. Common application extensibility architecture with embedded languages

Scenario. Assume C++ applications adopting Lua (Ierusalimschy, 2003) for extensibility. The Lua virtual machine is used as a library, being at runtime a separate execution system with its own call stack compared to the C++ runtime environment. Once an exception is thrown by Lua code the stack unwinding process is initiated by the Lua virtual machine in a way entirely transparent to the C++ runtime. In general, with the only exception of .NET that we examine in the Related Work section, the application and the embedded language runtimes are completely separated. Cross-language invocations concern either recursive invocations of the embedded language runtime (from application to extension) or invocation of application functions through the embedded language library mechanism (from extension to application).

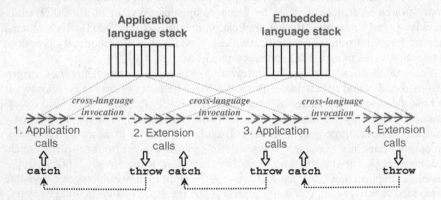

Fig. 2. Cross-language invocations and separate call stacks for applications extended by embedded languages

The scenario of intermixed calls, illustrated under Fig. 2, is very common in using embedded languages and is caused due to cross-language invocations. Calls made in the application code are pushed on the stack of the application language, while calls made in extension code are pushed on the stack of the embedded language.

The problem. For exceptions thrown by extension code (*throw* at point 2 and 4 of Fig. 2) it is not possible to define respective catch blocks inside application code. The reason is that embedded languages can trace catch blocks *only* in extension code. Thus *application developers are disabled to handle in application code exceptions thrown inside extensions*. The same applies in the opposite direction, hence *extension developers are disabled to handle in extension code exceptions thrown inside the application*. Although boost.Python (Boost, 2003) aimed to address the latter case for Python (Python, 2009), we discuss why the offered solution is suboptimal. In other highly popular languages such as Ruby (Flanagan & Matsumoto, 2008) and Perl (Wall et al., 2003) no support for cross-language exception handling is offered.

Contribution. To support cross-language exception handling we introduce a small set of additional features for embedded languages, discussing the required API amendments and respective runtime semantics. Our implementation was carried out as part of the Delta programming language (Savidis, 2010) whose source distribution is publicly available. Cross-language exceptions were exploited in the Delta language IDE which is implemented in C++ and incorporates many extensions written in the Delta language, including the source browser, debugger tree views and build tree. Such extensions may raise exceptions caught by native IDE code and vice versa.

2 Related Work

Cross-language exception handling in .NET languages. The .NET framework allows applications in one language to interoperate with applications or components written in another language. This is possible because due to presence of CIL (Common Intermediate Language) which is the .NET byte code format run by the CLI (the .NET virtual machine and ahead-of time compilation system).

This technical approach might constitute a universal solution to the problem of cross-exception handling only once all existing embedded and application languages conform to the .NET CLI specification. We consider that this is hardly ever possible, thus we need practical solutions applicable in combining diverse language runtimes.

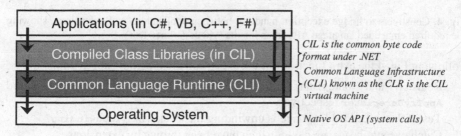

Fig. 3. Cross-language interoperation in the .NET framework due to CLI

boost.Python. All C++ exceptions are caught by the Python interpreter, then translated to Python values, if translators have been registered, and are propagated to Python user code. This suffers from the following issue: when no respective catch blocks exist in Python code a runtime error is issued, meaning any C++ catch blocks in the C++ call stack preceding the Python interpreter invocation are silently ignored.

3 Implementation Approach

throw ***in extension*** → catch ***in application*** Firstly, it requires the embedded language API to provide constructs for the definition of application-level catch blocks on exceptions thrown by invoked extension code. The key point is that such constructs will internally *notify the embedded language runtime for the presence of respective application-level catch blocks.* This way, if an exception is thrown by extension code, the embedded language exception system *is aware of all handlers* present within earlier application calls. In our implementation, such constructs are offered with the syntactic emulation of a special try-catch block (see Fig. 4) using a combination of macros and temporary definitions of the C++ language. The conditions for *if* statements commented as /* *never taken* */ are always *false*, but need to be evaluated; they are structured this way for syntax emulation purposes.

```
_try_          { invocation of extension code using the embedded language }
_catch_(e)     { handling logic for exception occurring in extension code   }

#define _try_                                                      \
        if (!ExceptionSystem::AppTryRegister())                    \
            {} /* never taken*/                                    \
        else
#define _catch_(e)                                                 \
        if (!ExceptionSystem::IsUnwinding())                       \
            ExceptionSystem::AppTryUnregister();                   \
        else if (ExceptionSystem::ValueType e = false)             \
            {} /* never taken*/                                    \
        else if (!ExceptionSystem::AppCatch(&e))                   \
            {} /* always taken*/                                   \
        else
#define _throw_(e)                                                 \
        if (ExceptionSystem::IsUnwinding())                        \
            ExceptionSystem::AppErrorThrowWhileUnwinding();        \
        else                                                       \
            ExceptionSystem::AppThrow(e)
```

Fig. 4. Constructs to bridge exception handling between applications and extensions, showing the required embedded language API amendments (via underlined text)

Following Fig. 4 the extra functions of the exception system *allow applications* to:

- Register and unregister try-catch blocks using **AppTryRegister** and **AppTryUnregister** respectively
- Test if the embedded language is unwinding the stack using **IsUnwinding**
- Catch exceptions via **AppCatch** which have been thrown by extensions
- Throw exceptions via **AppThrow** that can be caught by extensions

The previous functions suggest a *dynamic registration scheme*. More specifically, using a table driven scheme, catch blocks are traced and indexed statically during compilation. But then the embedded language runtime keeps information only for handlers appearing within extension code. Since the new type of catch blocks appears in application code, they are statically invisible to the embedded language, thus they should be registered manually and dynamically. Hence, an integrated dynamic registration scheme may be overall adopted for implementation consistency.

A detailed example is provided under Fig. 5 showing the steps involved during cross-language exception handling. More specifically, if an exception is thrown in extension code (by extension function *w*), the exception system will spot the presence of an application catch block registered earlier (by application function *F*). Then, it starts an unwinding process (*unwind-1*) which strips-off all call records (pops *w*, *q* and *p*) *until* the application library function *H* is met; the latter cannot be unwound by the embedded language since it resides on the application language call stack. In this case, the default function return sequence applies (*auto ret-1*). Following such a normal return, the control is back to the embedded language runtime which continues the unwinding process (*unwind-2*), eventually clearing the embedded language stack (pops *invoke_lib*, *g*, and *f*). This returns to the invocation of the embedded language runtime (*invoke_vm*, mentioned as *extension call* in the *_try_* block of *F*) at the application stack, at the end of the *_try_* block. Then, a control flow fallback leads directly to the user-defined *_catch_* block below which will handle the exception.

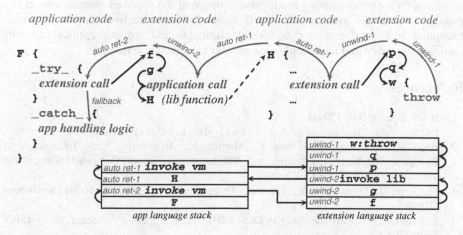

Fig. 5. Showing the call sequence, stacks and the unwinding process in a scenario throw *in extension* → catch *in application*

throw **in application** → catch **in extension** In boost.Python it is possible to throw native exceptions, i.e. at the application language level, which are handled by catch blocks in Python code. In this approach, *all native exceptions are filtered* and passed to the top Python interpreter instance on the call stack, *even if no user-defined handlers are found* in Python code. The latter disables the handling of exceptions by application catch blocks appearing below the Python interpreter in the native stack.

To address this issue we introduced a special _throw_ construct (see Fig. 5) which propagates exceptions *directly* to the embedded language runtime. This way, on the one hand we enable extensions handle exceptions thrown by the application, while on the other hand we ensure exceptions not handled by extensions are always propagated to the first available application _try_ / _catch_ block in the application call stack. In this context, we propose the following policy for application exceptions:

(i) All exception types being intrinsic, i.e. *private*, to applications should be managed only at the level of the exception system of the application language; and

(ii) All exception types being externalised, i.e. *public*, to extensions should be managed only at the level of the exception system of the embedded language through the special-purpose _try_, _catch_ and _throw_ constructs.

4 Conclusions

We discussed the need for cross-language exception handling in applications being extensible through embedded languages. To this end we propose to extend the exception system of the embedded language with extra special-purpose _try_, _catch_ and _throw_ constructs to be used *only within* application code in order to raise and handle cross-language exceptions (i.e., exceptions for which is allowed to cross the language boundaries). Such constructs internally cooperate with the stack unwinding mechanism of the embedded language. We avoided low-level details, putting emphasis on the programming model abstraction and the required amendments at the embedded language system API. We believe that the effective support of cross-exception handling for such dynamically extensible and growing applications will facilitate improved and more sophisticated fault tolerance practices.

References

1. Boost, boost.python (2003),
   http://www.boost.org/doc/libs/1_46_1/libs/python/doc/
2. Brito, P., de Lemos, R., Rubira, C., Martins, E.: Architecting Fault Tolerance with Exception Handling: Verification and Validation. Journal of Computer Science and Technology 24(2), 212–237 (2009)
3. Flanagan, D., Matsumoto, Y.: The Ruby Programming Language. O. Reilly, Sebastopol (2008)
4. Ierusalimschy, R.: Programming in Lua (2003), http://www.lua.org/pil/ ISBN 85-903798-1-7
5. Iliasov, A., Romanovsky, A.: Exception Handling in Coordination-Based Mobile Environments. COMPSAC (1), 341–350 (2005)
6. Loui, R.P.: In Praise of Scripting: Real Programming Pragmatism. IEEE Computer 41(7), 22–26 (2008)
7. Microsoft Corp., Cross-Language Interoperability, NET Framework 4, MSDN (2011)
8. Python language (2009), http://www.python.org/
9. Romanovsky, A.: Exception Handling in Component-Based System Development. In: COMPSAC, p. 580 (2001)
10. Savidis, A.: Delta Language (2010),
    http://www.ics.forth.gr/hci/files/plang/Delta/Delta.html

11. Wall, L., Christiansen, T., Orwant, J.: Programming Perl, 3rd edn. O'Reilly, Sebastopol (2000)
12. Walker, W., Koch, C., Gehrke, J., Demers, A.: Better Scripts, Better Games. Communications of the ACM 52(3), 42–47 (2009)
13. Xu, J., Romanovsky, A., Randell, B.: Concurrent Exception Handling and Resolution in Distributed Object Systems. IEEE Transactions on Parallel and Distributed Systems 11(10), 1019–1032 (2000)

Appendix

Extra Functions and Execution Control for Cross-Language Exception Handling

```
AppTryRegister {
    push a catch block marked as app-type
}

AppTryUnregister {
    assert that the top block is of app-type
    pop the top catch block
}

AppCatch(e) {
    copy the value of the internal exception var on e
    set is-unwinding = false
}

AppThrow(e) {
    copy the value of e on the internal exception var
    set is-unwinding = true
    Unwind()
}

Unwind() {
    if top catch block is of embedded-type and was registered by this execution loop then {
        set program counter to the respective catch block
        set is-unwinding = false
    }
    else { // either an embedded-type catch block of an outer execution loop or an app-type catch block
        set unwinding-is-postponed = true
        set exit-execution-loop = true
    }
}

ExecutionLoop() {
    set exit-execution-loop = false
    while not end of program and not exit-execution-loop do {
        if unwinding-is-postponed == true then {
            set unwinding-is-postponed = false
            Unwind()
        }
        else
            execute current instruction
    }
}
```

Guaranteeing Correct Evolution of Software Product Lines: Setting Up the Problem

Maurice H. ter Beek[1], Henry Muccini[2], and Patrizio Pelliccione[2]

[1] Istituto di Scienza e Tecnologie dell'Informazione "A. Faedo", CNR, Pisa, Italy
maurice.terbeek@isti.cnr.it
[2] Università degli studi dell'Aquila, Dipartimento di Informatica, L'Aquila, Italy
{henry.muccini,patrizio.pelliccione}@univaq.it

Abstract. The research question that we posed ourselves and which has led to this paper is: *how can we guarantee the correct functioning of products of an SPL when core components evolve?* This exploratory paper merely proposes an overview of a novel approach that, by extending and adapting assume-guarantee reasoning to evolving SPLs, guarantees the resilience against changes in the environment of products of an SPL. The idea is to selectively model check and test assume-guarantee properties on those SPL components affected by the changes.

1 Introduction

Product Line Engineering (PLE) aims to develop product families using a common platform and mass customization. Software Product Lines (SPLs) are part of an SPLE approach to develop, in a cost effective way, software-intensive products and systems that share an overall reference model of a product family [10]. Variety is achieved by identifying variation points as places in the Product Line Architecture (PLA) where specific decisions are reduced to the choice among several *features*, but the feature to be chosen for a particular product variant is left open (like optional, mandatory or alternative features). Variability management is the key aspect differentiating SPLE from 'conventional' software engineering.

Several approaches aim to verify correctness of SPLs (w.r.t. constraints and properties) and the conformance of products w.r.t. a family's variability [1, 2, 3, 4, 5, 6, 11]. SPLs however are subject to evolution, and their complexity makes it difficult to ensure the resilience of each product, i.e., the ability of the system to persistently perform in a trustworthy way even when facing changes. Apart from some work on regression testing, we are not aware of any existing approach on guaranteeing the correctness of products whose SPL or PLA is subject to evolution.

In this paper, we propose a step towards a solution to this problem by reusing, adapting, and combining currently available state-of-the-art techniques. More specifically, we intend to use assume-guarantee reasoning, which represents the environment as a set of properties that should be satisfied for it to interact correctly with the component. These properties are called *assumptions*, intended

E.A. Troubitsyna (Ed.): SERENE 2011, LNCS 6968, pp. 100–105, 2011.

to be the assumptions that the component makes on the environment. If these assumptions are satisfied by the environment, then components behaving in this environment will typically satisfy other properties, called *guarantees*. By appropriately combining the set of assume and guarantee properties, it is possible to prove the correctness of an entire system without actually constructing it.

Our idea is then to annotate components of a PLA with pairs of assumption and guarantee properties, considering a component's environment as the composition of the remaining components, thus enabling assume-guarantee compositional verification to deal with evolution. The idea underlying *compositional verification* is to decompose a system specification into properties that describe the behavior of a system's subset, to check these properties locally, and to deduce from the local checks that the complete system satisfies the overall specification. The main advantage is that one never has to compose all subsystems, thus avoiding the state explosion problem. On the downside, however, checking local properties over subsystems in general does not imply the correctness of the entire system, due to the existence of mutual dependencies among components. More precisely, a single component cannot be considered in isolation, but as behaving and interacting with its environment (i.e., the rest of the system).

In the future, we moreover intend to adapt the assume-guarantee testing approach of [9] to evolving PLAs, to complement architecture-level verification with code-level testing. This would allow one to guarantee the resilience of products of an evolving SPL by selectively model checking and testing affected components.

2   Problem Setting

Figure 1 shows the context of our work. During *domain engineering*, the whole SPL is taken into account: commonality and variability of the software family are analyzed in order to identify both the common components and the variation points. Domain assets are produced during the domain engineering phase to describe the requirements, architecture and tests of the entire family of applications. This current work focusses on architectural assets, thus focussing on the Product Line Architecture (PLA) of the SPL.

During the *application engineering* phase, instead, a decision process (more or less explicit) is then typically employed to decide which optional, mandatory or alternative features need to be selected in order to produce a certain software system from the product line. The application level product reflects development constraints and decisions made during the decision process, and its architecture is denoted as a Product Architecture (PA).

In this context, we consider the problem of guaranteeing the correct evolution of the SPL upon changes in components of the PLA. Figure 2 illustrates our problem setting by means of the following model problem,[1] which is due to Paul Clements et al. at the SEI:

---

[1] A model problem has been defined as a problem that, if solved, would result in a significant decrease in project resources devoted to testing and analysis and/or a significant increase in system quality given an expenditure level.

Domain Engineering

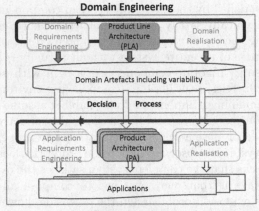

Application Engineering

Fig. 1. Software Product Line

"I run the same software in different kinds of helicopters. When the software in a helicopter powers up, it checks a hardware register to see what kind of helicopter it is and starts behaving appropriately for that kind of helicopter. When I make a change to the software, I would like to flight test it only on one helicopter, and prove or (more likely, assert with high confidence) that it will run correctly on the other helicopters. I know I can't achieve this for all changes, but I would like to do it where possible."

In the context of SPLE, this model problem can be re-phrased as follows: assuming various products (helicopters) have been derived from an SPL — which have moreover been formally certified — what can be concluded for new products obtained from the SPL by modifying one or more core components?

We assume that all products of the product line must be guaranteed to conform to a specific standard. Furthermore, we assume that there is a policy according to which any change to a core component requires all products containing that core component to be rebuilt. The question is whether it is necessary to re-validate all the products of the SPL or whether a subset can suffice. For instance, in the aforementioned model problem, when we change the software of the helicopter line, such as installing a new kind of radio across the fleet, we would like to flight-test it only on one helicopter, and prove or — more likely — assert with high confidence, that it will run correctly on the other helicopters. While it is impossible to achieve this for all kind of changes, we would like to pinpoint the conditions under which this is feasible.

Based on this problem, our research question becomes: *After a component C of a product is modified and the resulting product is guaranteed correct, what can be guaranteed for the other products of the product family that also contain C?* Different scenarios can be distinguished, and in this paper we sketch a solution for modifications resulting from changing, adding or removing such components.

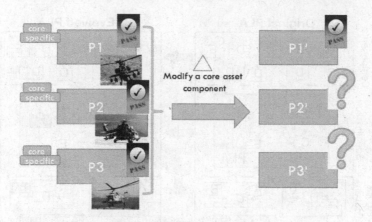

Fig. 2. Helicopter problem

3   Proposed Solution

While much work has been proposed for analyzing and testing SPLs, we are not aware of any approach for guaranteeing their products' correctness upon evolving components. Techniques for the formal verification of SPLs require a complete re-run of the verification after evolution, whereas regression testing of SPLs is not able to provide many guarantees for success. Finally, while compositional verification techniques are well suited to be applied to evolving systems, we know of no approach that is directly applicable to SPLs.

Figure 3 illustrates our idea of formally verifying that modifications to a component that is part of a PLA do not have unforeseen consequences. Each component is enriched with assume and guarantee properties, by properly calculating assumptions characterizing exactly those environments in which the component satisfies its required properties. In [8], this is done by means of the automatic generation of assumptions implemented in the LTS Analyser. We intend to adapt and extend this approach to consider evolving product line architectures.

When the PLA of the product family of interest is specified, assume and guarantee properties are contextualized: given a component C with its assume-guarantee pair, denoted by $C(a, g)$, its environment becomes the subarchitecture connected with C. What is challenging in PLA-based assume-guarantee reasoning is that the reasoning is performed on the PLA, including all variation points, rather than on each single concrete product, which means that the assumptions need to be calculated considering the variability management of components. In other words, once a component evolves this modification will potentially impact several PAs, namely each PA containing the modified component.

For instance, when a component B evolves in a component B' (see Fig. 3), the assumption and guarantee pairs of B and B' must be checked:

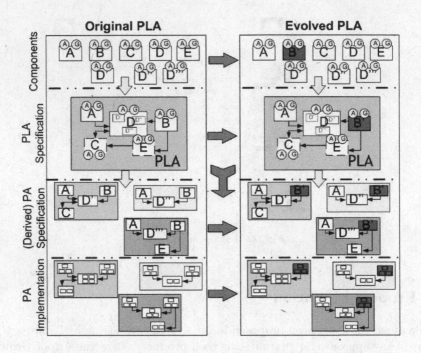

Fig. 3. The original SPL (left) and the evolved one (right)

- If $B'(a,g)$ exactly matches $B(a,g)$, then B' can safely substitute B.

- If $B'(a,g)$ is less restrictive than $B(a,g)$, in the sense that the assumption of B' is less restrictive than that of B, then it is possible to restrict the behavior of B' to that of B by means of suitable wrappers, thus forcing B' to behave as requested by the assumption of B.

- If B' is not able to behave as expected, according to $B(a,g)$, then assume-guarantee reasoning can be used to understand the effect of the evolution. In fact, since the components composing the PLA are organized in a chain of assumptions and guarantees (through which the composition of the PLA is realized) it is possible to understand the effect of changes in a component.

Components may moreover represent various alternatives, all of which share some basic properties while they differ in specific details. Formally, we intend to adapt the approach of [1] and define a component C as a Modal Transition System (MTS) that represents its n different variants $C_1, \ldots, C_n$ as Labelled Transition Systems (LTSs). Then there exist assume and guarantee properties such that for all $i \in \{1, \ldots, n\}$, $C_i(a,g)$ either exactly matches, or is less restrictive than, $C(a,g)$. When C is composed with another component D according to $C(a,g)$, then also each C_i can be composed with D. Obviously, it may be the case that $C(a,g)$ does not provide enough guarantees to D, but that there exists a component C_j, with $i \neq j \in \{1, \ldots, n\}$, which does provide the required guarantees, in which case only this variant C_j can be composed with D.

As soon as the PLA components are defined as MTSs, the MTS Analyser [7] — a tool built on top of LTSA to deal with MTSs rather than LTSs — can be used in order to properly adapt the approach in [1] to PLAs. The attained result is that only parts of the assume-guarantee chain need to be re-verified, thus minimizing the amount of re-verification needed after a component evolves.

4   Conclusion and Future Work

In this paper, we presented a preliminary approach to guarantee the resilience of products of an SPL when facing changes in some of its components. The approach is based on assume-guarantee reasoning and permits to selectively model check and test only those components affected by evolution. Our expectation is that this considerably reduce the effort needed to re-analyze products of an SPL upon a change in that SPL. As future work we plan to implement the overall approach and to experiment its effectiveness by applying it to several case studies.

References

1. Asirelli, P., ter Beek, M.H., Fantechi, A., Gnesi, S.: Formal description of variability in product families. In: SPLC 2011. Springer, Heidelberg (to appear, 2011)
2. Asirelli, P., ter Beek, M.H., Fantechi, A., Gnesi, S., Mazzanti, F.: Design and validation of variability in product lines. In: Proceedings PLEASE 2011, pp. 25–30. ACM, New York (2011)
3. Classen, A., Heymans, P., Schobbens, P.-Y., Legay, A.: Symbolic model checking of software product lines. In: Proceedings ICSE 2011, pp. 321–330. ACM, New York (2011)
4. Classen, A., Heymans, P., Schobbens, P.-Y., Legay, A., Raskin, J.-F.: Model checking lots of systems: efficient verification of temporal properties in software product lines. In: Proceedings ICSE 2010, pp. 335–344. ACM, New York (2010)
5. da Mota Silveira Neto, P.A.: A Regression Testing Approach for Software Product Lines Architectures: Selecting an efficient and effective set of test cases. LAP (2010)
6. da Mota Silveira Neto, P.A., do Carmo Machado, I., McGregor, J.D., de Almeida, E.S., de Lemos Meira, S.R.: A systematic mapping study of software product lines testing. Inf. Softw. Technol. 53(5), 407–423 (2011)
7. D'Ippolito, N., Fischbein, D., Chechik, M., Uchitel, S.: MTSA: The modal transition system analyser. In: Proceedings ASE 2008, pp. 475–476. IEEE, Los Alamitos (2008)
8. Giannakopoulou, D., Pasareanu, C., Barringer, H.: Component verification with automatically generated assumptions. Autom. Softw. Eng. 12(3), 297–320 (2005)
9. Giannakopoulou, D., Pasareanu, C., Blundell, C.: Assume-guarantee testing for software components. IET Softw. 2(6), 547–562 (2008)
10. Pohl, K., Böckle, G., van der Linden, F.: Software Product Line Engineering: Foundations, Principles, and Techniques. Springer, Heidelberg (2005)
11. Thüm, T., Schaefer, I., Kuhlemann, M., Apel, S.: Proof composition for deductive verification of software product lines. In: Proceedings VAST 2011 (to appear, 2011)

Idealized Fault-Tolerant Components
in Requirements Engineering

Sadaf Mustafiz and Jörg Kienzle

School of Computer Science, McGill University
Montreal, Quebec, Canada
sadaf@cs.mcgill.ca, joerg.kienzle@mcgill.ca

Abstract. We have previously proposed a requirements development process, DREP, for dependable systems. In this paper, we draw a parallel between the notions defined in DREP and the elements of idealized fault-tolerant components (IFTCs). We show how the key ideas of IFTCs can be re-interpreted at the requirements engineering level and mapped to DREP concepts.

1 Introduction

The *Idealized Fault-Tolerant Component* (IFTC) has been proposed in the 80's as a basis for designing dependable systems [1]. An IFTC provides reliable service upon request, and signals abnormal situations using exceptions. To make modular decomposition possible, IFTCs can form hierarchies. Together with rigorous exception handling and propagation policies, IFTCs simplify the structuring and execution of complex systems. This has been proven in the past 40 years by many concrete realizations of the IFTC concept, which were successfully used to design and implement dependable software for safety and mission-critical systems.

Like a lot of other work in the dependable system domain, the IFTC focuses on the *design* of fault-tolerant systems. The early stages of development are often not given due attention, and it has been reported that a major concern in the development of dependable systems are flawed system specifications due to *incomplete requirements* [2]. Motivated by that fact, recent research has focused on extending standard requirements elicitation processes to address potential abnormal situations that can interrupt normal system interaction at run-time. In particular, we have proposed the Dependability-Oriented Requirements Engineering Process (DREP) [3,4,5] that extends traditional use case-driven requirements elicitation to consider reliability and safety concerns. After the discovery of normal system behaviour by means of use cases, the developer is lead to explore exceptional situations arising in the environment that change the context in which the system operates and service-related exceptional situations that threaten to fail user goals. The process requires the developer to specify means that detect such situations, and to define the recovery measures that attempt to put the system in a reliable and safe state. The process is iterative, and refinements are carried out, if necessary, to achieve desired quality levels.

We realized that many of our extensions, such as handler use cases and exceptional outcomes, are similar to ideas proposed in the context of IFTCs, but re-interpreted in the context of requirements engineering. Motivated by this observation, we present in

E.A. Troubitsyna (Ed.): SERENE 2011, LNCS 6968, pp. 106–112, 2011.

this paper a detailed comparison between concepts defined in the realm of system design using IFTCs and concepts we proposed to be used when eliciting and specifying behavioural requirements for dependable systems. The ultimate goal of this exercise is to be able to transfer insights gained during the last 40 years in system design using IFTCs to the requirements phase.

2   Idealized Fault-Tolerant Components in DREP

In 1981, Anderson and Lee introduced a methodology and framework for designing dependable systems [1]. The basic building block of their framework is the *Idealized Fault-Tolerant Component* (IFTC)[1]. An IFTC is a *design or implementation module* that encapsulates the structure and behaviour to implement one or several *well-defined system functionalities*. It can encapsulate *hardware devices*, but also *software components*. It has a *well-defined interface* that defines the services it provides to other components in the system. As a result of the activity of a system, the IFTC receives requests to provide service. By means of normal activity, the IFTC will endeavour to provide the requested service according to its specification.

The IFTC framework allows hierarchical structuring of components: an IFTC can contain one or several sub-components. However, containment must be strictly nested, i.e. a sub-component can be part of one parent component only. At run-time, an IFTC providing a requested service can rely on sub-components to do. In this case, the IFTC sends itself requests to its sub-components. If these sub-components respond normally, and if no errors are detected by the IFTC itself, then a "legal service request" produces a "normal response". The interface of an IFTC clearly specifies the service requests it accepts as well as the normal responses it generates as a result.

Fig. 1, adapted from [1], illustrates the behaviour of an IFTC at run-time and how it switches from the normal processing mode to an error processing mode. A service request is received from the outside (request arrow coming into the IFTC from the top left). The IFTC can make requests to sub-components (request arrow leaving the IFTC on the bottom left). The figure also illustrates that not only local exceptions can activate the error processing mode, but also interface or failure exceptions signalled by a sub-component.

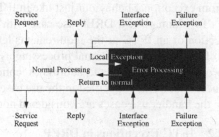

Fig. 1. IFTC

2.1   IFTCs in DREP

At the highest level of abstraction, the system under development in requirements engineering is viewed as a black-box and considered to have no internal components.

---

[1] The original publication actually used the name *Ideal Fault-Tolerant Component*, but all subsequent publications we know of used the word *Idealized*.

The system provides a set of services that actors in the environment can request to be provided by interacting with the system.

In DREP, each service is described by a use case. What DREP adds over other use case-based approaches is that it leads the requirements engineer to explore exceptional situations that interfere with normal service provision. Very often, an exceptional situation that arises while a system is providing service makes it impossible for the system to complete the task at hand. For instance, the system might be relying on an external device to perform a certain action, but the device is malfunctioning. The *reliability* of the service, or maybe even system and user *safety* are at stake. When the original service cannot be provided, a dependable system should strive to handle the current situation. In DREP, any special interaction with the environment that might be necessary is specified in *handler use cases*.

Handler use cases are classified as *safety* or *reliability* handers. The former do not attempt to continue the current service provision, but rather perform actions to ensure system and user safety. The latter try to adapt to the current situation and provide the service in a different way. If that is not possible, providing partial service is also an option. However, since any observable deviation from a service's specification would be considered a failure, DREP advocates that in addition to defining the success outcome(s) for each service, a specification should also include definitions of **degraded outcomes** for each service.

Just like in the original IFTC framework where IFTCs can contain sub-components, use cases in requirements engineering can contain subfunction-level use cases. So although at the requirements level the system is viewed as a black box without internal components, the description of the services that the system provides can be broken up into smaller pieces, creating a hierarchy of use cases. In DREP, each use case is also associated with the handler use cases that describe how to cope with exceptions that might occur while executing the use case. This clearly separates normal behaviour from exceptional behaviour just like in IFTCs.

To summarize, a DREP use case with its associated handler use cases is the projection of an IFTC at the requirements level. It accepts service requests and attempts to fulfill them using normal processing as described in the use case, and defines error processing interactions to address exceptional situations in the handler use cases. The success outcomes specified in the use case, as well as the degraded outcomes specified in the handler use cases are considered normal responses of the IFTC.

2.2   IFTC Exceptions in DREP

The IFTC framework defines three different kinds of exceptions: *interface exceptions*, *local exceptions* and *failure exceptions*. Interface and failure exceptions cross the boundary of an IFTC, whereas local exceptions move the component from normal processing to exceptional processing mode. DREP defines two types of exceptions at the requirements level: *context-affecting exceptions*, i.e. exceptional situations arising in the environment that change the context in which the system operates, and *service-related exceptions*, i.e. situations that threaten to fail user goals.

Local Exceptions. *Service-related exceptions* occur while the system is fulfilling a specific service request, performing interactions with the environment as specified in a use case. When a service-related exception occurs, the appropriate handler use case attached to the use case is executed in order to attempt to address the situation. Therefore, local exceptions in IFTCs correspond to service-related exceptions in DREP.

Failure Exceptions. According to the original definition, a failure exception in the IFTC framework must be signalled by a component when a request cannot be serviced despite internal error processing. DREP requires handler use cases to clearly specify the outcome of the handling activity, which can be either *success*, *degraded* or *failure*. As discussed above, a reliability handler that manages to provide the requested service in an alternate way declares a success outcome. A partial service provision, signalled by a degraded outcome, is also considered a normal response of the system. However, reliability handers that cannot achieve partial service or safety handlers that interrupt service provision for the sake of safety must signal a failure outcome to the enclosing use case (or the environment). Hence, failure exceptions in IFTCs map to DREP failure outcomes.

The debate about whether or not in case of a component failure the state of the component is restored to the state that was valid before the last request was received is not relevant at the level of abstraction of DREP. At the requirements level, the actual state that the system needs to keep in order to provide the specified services is not defined. What is relevant, however, is whether or not the system, given the current situation, is capable of fulfilling its normal functionality in case of future service requests. DREP addresses this issue by providing different operation modes as explained in the following subsection.

Interface Exceptions. In the IFTC framework, a component signals an interface exception if it receives an illegal or malformed service request. In DREP this would correspond to an actor in the environment requesting a service from the system that it currently does not provide. This situation may arise because DREP guides the requirement engineer to evaluate the effects of exceptional situations on future service provision of the system. In the case where the current situation might affect future services, the system should switch from its normal mode of operation into a different mode in which the services that are affected by the problem are not offered anymore.

Context-affecting exceptions are most likely to cause a mode switch, since they represent events that happen in the environment that change the context in which the system operates. In a different context it often makes no sense to continue to offer the normal system services. In fact, doing so might be impossible or even dangerous.

In some cases, service-related exception also cause mode switches. For instance, a service-related exception that arises when interacting with a device, especially when occurring repeatedly, can be a sign that the device has failed permanently. If the device was vital for the system to achieve the desired functionality, the requested service cannot be offered anymore in the future.

When an actor in the environment requests a service from the system that is not provided in the current mode, the system replies with an (exceptional) output message to notify the actor that the service cannot be provided at this time. An IFTC interface exception therefore corresponds to such an output message.

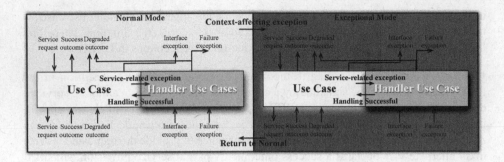

Fig. 2. Integration of DREP Concepts with the IFTC Framework

2.3 Mapping Summary

- An **IFTC** takes the form of a **service** in DREP.
- Normal processing is described by a **use case** in DREP.
- Error processing is described by multiple **handler use cases** in DREP.
- A **service request** refers to a **request for a service made by an actor**.
- A **normal reponse** in DREP refers to the **success and degraded service outcomes** of the use case and attached handler use cases.
- A **local exception** corresponds to a **service-related exception** in DREP, which invokes the corresponding handler use case.
- A **interface exception** happens when an actor issues an **invalid service request**, i.e. a request for a service that is not offered in the current operating mode.
- A **failure exception** corresponds to a **failure outcome** signalled by a use case or handler use case while providing a requested service.

Figure 2 illustrates how the DREP concepts fit with the IFTC framework. One IFTC that represents the *System* includes IFTCs that represent the services, the behaviour of which is described by use cases and handler use cases. Although not shown in the figure, a service can be decomposed further into sub-services.

3 Related Work

Rubira et al. [6] present an approach that incorporates exceptional behavior in component-based software development by extending the Catalysis method. The components are structured as IFTCs. Guerra et al. [7] proposes a similar approach for structuring exception handling for fault-tolerant component-based systems.

Bucchiarone et al. [8] focus on validation of component-based systems. The approach begins by addressing exceptions in requirements, and structures and classifies the requirements and exceptional behaviour according to the IFTC model. The elements are mapped on to construct a fault-tolerant architecture, and then testing is carried to ensure that the design conforms to the fault-tolerant requirements.

Brito et al. [9] propose MDCE+, a method for modelling exceptional behaviour in component-based software development starting from requirements to testing. Similar to our ideas in DREP, the method also aims to discover exceptions and define handlers early on in the development lifecycle. At the requirements phase of the MDCE+ method, the activities are structured and the exceptions are specified according to the IFTC model.

4   Discussion

This paper shows how the key ideas of the IFTC framework that was developed to structure the *architecture* and *design* of dependable systems can be re-interpreted and applied in the context of *requirements engineering*. We discuss how the IFTC constructs are related to the dependability concepts we introduced in our dependability-oriented requirements engineering process DREP. In particular, we relate IFTCs, requests and responses, local exceptions, interface exceptions and failure exceptions to use cases and handlers, service requests, normal outcomes, degraded outcomes, failure outcomes, service-related exceptions, context-affecting exceptions and modes of operations.

The benefit of doing this exercise is that good practices gained during years of experience designing fault-tolerant systems using IFTCs can be applied at the requirements engineering level.

For instance, good exception handling practice dictates that every IFTC must provide handlers for all local exceptions as well as interface and failure exceptions of contained IFTCs. At the requirements level, this translates to every service providing handlers for service-related exceptions that could arise during service provision, as well as defining handlers for failure outcomes of included subfunction-level use cases.

Another well-known principle in the world of IFTCs is that exceptions of a subcomponent should never be propagated as is to a higher level, since a higher level should not be exposed to the internal design details of lower levels. Instead, a lower level exception that cannot be handled should be translated into a meaningful exception for the higher level that does not reveal unnecessary implementation details. The same principle could be applied at the requirements engineering level.

Finally, we believe that many ideas used in concrete realizations of the IFTC framework can also be applied at the requirements engineering level. For instance, recovery blocks [10] achieve fault tolerance by executing a primary algorithm first. In case of a failure, the state of the component is rolled back and and a backup algorithm is executed. Similarly, at the requirements level, a service can first try and achieve the specified functionality by executing the primary subfunction-level use case, and in case of a failure outcome switch to a backup use case.

References

1. Anderson, T., Lee, P.: Fault Tolerance - Principles and Practice. Prentice-Hall, Englewood Cliffs (1981)
2. Davis, A.M.: Software requirements: objects, functions, and states. Prentice-Hall, Englewood Cliffs (1993)
3. Shui, A., Mustafiz, S., Kienzle, J., Dony, C.: Exceptional use cases. In: Briand, L.C., Williams, C. (eds.) MoDELS 2005. LNCS, vol. 3713, pp. 568–583. Springer, Heidelberg (2005)

4. Mustafiz, S., Kienzle, J.: DREP: A requirements engineering process for dependable reactive systems. In: Butler, M., Jones, C., Romanovsky, A., Troubitsyna, E. (eds.) Methods, Models and Tools for Fault Tolerance. LNCS, vol. 5454, pp. 220–250. Springer, Heidelberg (2009)
5. Mustafiz, S.: Dependability-Oriented Model-Driven Requirements Engineering for Reactive Systems. PhD thesis, McGill University (2010)
6. Rubira, C.M.F., de Lemos, R., Ferreira, G.R.M., Filho, F.C.: Exception handling in the development of dependable component-based systems. Software, Practice & Experience 35, 195–236 (2005)
7. de Lemos, R., de Castro Guerra, P.A., Rubira, C.M.F.: A fault-tolerant architectural approach for dependable systems. IEEE Software 23, 80–87 (2006)
8. Bucchiarone, A., Muccini, H., Pelliccione, P.: Architecting fault-tolerant component-based systems: from requirements to testing. Electr. Notes Theor. Comput. Sci. 168, 77–90 (2007)
9. da S. Brito, P.H., Rocha, C.R., Filho, F.C., Martins, E., Rubira, C.M.F.: A method for modeling and testing exceptions in component-based software development. In: Maziero, C.A., Gabriel Silva, J., Andrade, A.M.S., de Assis Silva, F.M. (eds.) LADC 2005. LNCS, vol. 3747, pp. 61–79. Springer, Heidelberg (2005)
10. Horning, J.J., Lauer, H.C., Melliar-Smith, P.M., Randell, B.: A program structure for error detection and recovery. In: Goos, G., Hartmanis, F., (eds.) Operating Systems. LNCS, vol. 16, pp. 172–187. Springer, Heidelberg (1974)

Predictability and Evolution in Resilient Systems

Ivica Crnkovic

School of Innovation, Design and Engineering, Mälardalen University, Västerås, Sweden
ivica.crnkovic@mdh.se

Abstract. This paper gives a short overview of the talk related to the challenges in software development of resilient systems. The challenges come of the resilience characteristic as such; it a system emerging lifecycle property, neither directly measurable nor computable. While software is an essential part of a system, its analysis not enough for determining the system resilience. The talk will discuss about system resilience reasoning, its limitations, and possible approaches in the software design that include resilience analysis.

1 System Resilience, Robustness and Sustainability

In the last decade the importance of resilience as an intrinsic system attribute is arising in every domain of system and software development. Resilience is an attribute often related to robustness, and survivability (and by this dependability) from one side, and sustainability from other side [1]. Robustness and dependability are the important properties of many systems. Such systems are traditionally analyzed and designed in a top-down manner by defining the system boundaries, requirements and the system specifications with a goal to achieve predictable behavior in predictable (though not necessary known) situations. The main challenges and the main problems in these types of systems occur due to unforeseen events. For this reasons one important element of the quality assurance process is to ensure that such events do not happen, or the probability they can happen is minimized. The sustainability attribute on the other hand is the capacity to endure, related to resource constraints and maintenance of status quo of the external environment (i.e low negative impact in short and long terms). These characteristics are complementary and not necessary compatible, likely to require trade-off analysis to find a satisfactory balance. The resilience property should embodies both these properties the resilient systems should provide their services in a trustworthy and sustainable way even when the circumstances change; this may include drastic changes. More and more systems are required to perform in this way. The systems became more complex, more interactive, being integrated with another systems, and are continuously evolving. For this reason new characteristics of such systems become important; these systems must be adaptive and self-organizing. They are evolving, as the external environment is evolving.

Since software as an immanent part of most of the systems, the software development has to include the same concerns of the system resilience design, though the method can be specific for software. This talk will focus on two different software development aspects for resilient systems bottom-up and top-down approaches; a) on the software and system attributes composability, and b) software evolution analysis.

E.A. Troubitsyna (Ed.): SERENE 2011, LNCS 6968, pp. 113–114, 2011.

2   Compositions of System and Software Attributes Limitations

The modern paradigm of software development includes (re)use and possible adaptation of the existing software, either in a form of components, services, underlying platforms and infrastructure, or even complete software systems. This calls for design methods that use bottom-up approach, and composition of the existing components. The efficient development requires not only efficient mechanism for composition of functional properties, but also extrafunctional properties. In the concrete case, we can sate a question: Can we, and under which assumptions, reason about resilience composition? This talk will present different types of extrafuctional properties with respect to their compositions and composition assumptions [2,3].

3   Software Evolvability Analysis and Resilience Property

A complementary approach to composability is a top-down approach analysis of a system in respect to resilience. Since resilience is a complex property, it depends on a number of subproperties that may be easier analyzed and in some cases measured. The question is which are these subproperties. In this talk we shall present a method and a model for analyzing software evolvability [4] , one of the subcharacteristics of resilience. The example, based on industrial case studies [5], shows a method for evolvability assessment of software architecture. The method includes identification of subcharacteristics of evolvability and the assessment process, based on quantitative and qualitative methods, The example can be used as a starting point for resilience assessment.

References

1. Fiksel, J.: Designing resilient, sustainable systems. Environmental Science and Technology 37(23), 5330–5339 (2003)
2. Crnkovic, I., Sentilles, S., Vulgarakis, A., Chaudron, M.R.: A classification framework for software component models. IEEE Transactions on Software Engineering (2010) (in press)
3. Crnković, I., Larsson, M., Preiss, O.: Concerning predictability in dependable component-based systems: Classification of quality attributes. In: de Lemos, R., Gacek, C., Romanovsky, A. (eds.) Architecting Dependable Systems III. LNCS, vol. 3549, pp. 257–278. Springer, Heidelberg (2005)
4. Breivold, H.P., Crnkovic, I., Larsson, M.: A systematic review of software architecture evolution research, information and software technology. Information and Software Technology (June 2011) (in press)
5. Pei-Breivold, H., Crnkovic, I., Eriksson, P.: Analyzing software evolvability. In: IEEE International Computer Software and Applications Conference, COMPSAC 2008 (July 2008)

Self-organising Pervasive Ecosystems:
A Crowd Evacuation Example*

Sara Montagna, Mirko Viroli, Matteo Risoldi, Danilo Pianini,
and Giovanna Di Marzo Serugendo

[1] Università di Bologna – Via Venezia 52, IT-47521 Cesena
{sara.montagna,mirko.viroli,danilo.pianini}@unibo.it
[2] Université de Genève – Rte. de Drize 7, CH-1227 Carouge
{matteo.risoldi,giovanna.dimarzo}@unige.ch

Abstract. The dynamics of pervasive ecosystems are typically highly unpredictable, and therefore self-organising approaches are often exploited to make their applications resilient to changes and failures. The SAPERE approach we illustrate in this paper aims at addressing this issue by taking inspiration from natural ecosystems, which are regulated by a limited set of "laws" evolving the population of individuals in a self-organising way. Analogously, in our approach, a set of so-called *eco-laws* coordinate the individuals of the pervasive computing system (humans, devices, signals), in a way that is shown to be expressive enough to model and implement interesting real-life scenarios. We exemplify the proposed framework discussing a crowd evacuation application, tuning and validating it by simulation.

Keywords: pervasive computing, software ecosystems, self-adaptation, self-organisation.

1 Introduction

The increasing evolution of pervasive computing is promoting the emergence of decentralised infrastructures for pervasive services. These include traditional services with dynamic and autonomous context adaptation (e.g., public displays showing information tailored to bystanders), as well as innovative services for better interacting with the physical world (e.g., people coordinating through their PDAs). Such scenarios feature a number of diverse sensing devices, personal and public displays, personal mobile devices, and humans, all of which are dynamically engaged in flexible coordinated activities and have to account for resilience when conditions change. Recent proposals in the area of coordination models and middlewares for pervasive computing scenarios try to account for issues related to spatiality [12,14], spontaneous and opportunistic coordination [1,8], self-adaptation and self-management [20]; however, most works propose

---

* This work has been supported by the EU-FP7-FET Proactive project SAPERE—
Self-aware Pervasive Service Ecosystems, under contract no.256873.

E.A. Troubitsyna (Ed.): SERENE 2011, LNCS 6968, pp. 115–129, 2011.

ad-hoc solutions to specific problems in specific areas, and lack generality. The SAPERE project ("Self-adaptive Pervasive Service Ecosystems") addresses the above issues in a uniform way by means of a truly self-adaptive pervasive substrate; this is a space bringing to life an ecosystem of individuals, namely, of pervasive services, devices, and humans. These are coordinated in a self-organising way by basic laws (called *eco-laws*), which evolve the population of individuals in the system, thus modelling diverse mechanisms of coordination, communication, and interaction. Technically, such eco-laws are structured as sort of chemical reactions, working on the "interface annotation" of components residing in neighbouring localities—called *LSA* (Live Semantic Annotation).

A notable application of the proposed approach is in resilient crowd steering applications, in which a crowd is guided in a pervasive computing scenario depending on unforeseen events, such as the occurrence of critical events (i.e. alarms) and the dynamic formation of jams. We exemplify the approach in a crowd evacuation scenario, providing its set of eco-laws and validating it via simulation.

The remainder of the paper is organised as follows. In Section 2 we give a brief overview of the SAPERE approach and general architecture. Section 3 examines in detail the language for eco-laws. Section 4 describes the concrete example of crowd evacuation application, which is then validated in Section 5 by simulation. Related work and conclusions wrap up the article.

2   Architecture

The SAPERE approach is inspired by the mechanism of chemical reactions [24]. The basic idea of the framework is to model all the components in the ecosystem in a uniform way, whether they are humans perceiving/acting over the system directly or through their PDAs, pervasive devices (e.g., displays or sensors), or software services. They are all seen as external components (i.e., agents), reifying their relevant interface/behavioural/configuration information in terms of an associated semantic representation called *Live Semantic Annotation* (LSA). To account for dynamic scenarios and for continuous holistic adaptation and resilience, we make LSAs capable of reflecting the current situation and context of the component they describe. As soon as a component enters the ecosystem, its LSA is automatically created and injected in the SAPERE substrate, which is a shared space where all LSAs live and interact. Topologically, this shared space is structured as a network of LSA-spaces spread in the pervasive computing system and containing the LSAs of the associated components, each hosted by a *node* of the SAPERE infrastructure. Proximity of two LSA-spaces implies direct communication abilities.

Each LSA-space embeds the basic laws of the ecosystem, called *eco-laws*, which rule the activities of the system by evolving the population of LSAs. They define the policies to rule *reactions* among LSAs, enforcing coordination of data and services. LSAs (and thus their associated data and services) are like chemical reagents in an ecology in which interactions and composition occur via chemical-like reactions featuring pattern-matching between LSAs. Such reactions *(i)* change the status of LSAs depending on the context (e.g., a display showing information on the nearest exit only when a bystander needs it), *(ii)* produce new LSAs (e.g., a composite service coordinating the execution of atomic service components), or *(iii)* diffuse LSAs to nearby nodes (e.g., propagating an alarm, or creating a gradient data structure).

Coordination, adaptivity, and resilience in the SAPERE framework are not bound by the capability of individual components, but rather emerge in the overall dynamics of the ecosystem. Changes in the system (and changes in its components, as reflected by dynamic changes in their LSAs) result in the firing of eco-laws, possibly leading to the creation/removal/modification of other LSAs. Thus, the SAPERE architecture promotes adaptivity and coordination not by enacting resilience at the level of components, but rather promoting a sort of "systemic resilience".

3   Eco-law Language

We here present the eco-law language informally. Although in the SAPERE framework LSAs are *semantic* annotations, expressing information with same expressiveness of standard frameworks like RDF, we here consider a simplified notation without affecting the expressiveness of the self-organisation patterns we describe. Namely, an LSA is simply modelled as a tuple $\langle v_1, \ldots, v_n \rangle$ (ordered sequence) of typed values, which could be for example numbers, strings or structured data. For instance, $\langle 10001, \texttt{sensor}, \texttt{temperature}, 28 \rangle$ could represent the LSA with identifier 10001, injected by a sensor which currently measures temperature 28. In writing LSAs and eco-laws, we shall use `typetext` font for concrete values, and *italics* for variables.

An eco-law is a chemical-resembling reaction working over patterns of LSAs. An LSA pattern P is basically an LSA which may have some variable in place of one or more arguments of a tuple, and an LSA L is said to match the pattern P if there exists a substitution of variables which applied to P gives L. An eco-law is hence of the kind $P_1, \ldots, P_n \overset{r}{\mapsto} P'_1, \ldots, P'_m$, where: *(i)* the left-hand side (reagents) specifies patterns that should match LSAs $L_1, \ldots, L_n$ to be extracted from the LSA-space; *(ii)* the right-hand side (products) specifies patterns of LSAs which are accordingly to be inserted back in the LSA-space (after applying substitutions found when extracting reagents, as in standard logic-based rule approaches); and *(iii)* rate r is a numerical positive value indicating the average frequency at which the eco-law is to be fired—namely, we model execution of the eco-law as a CTMC transition with Markovian rate (average frequency) r. As a simple example, the eco-law $\langle 10001, \texttt{a}, 10 \rangle \overset{1.0}{\mapsto} \langle 10001, \texttt{a}, 11 \rangle$ fires when $\langle 10001, \texttt{a}, 10 \rangle$ is found in a space, its effect is to increment value 10 to 11, and the rate of its application is 1.0—an average of once per time unit. A more general eco-law $\langle id, \texttt{a}, 10 \rangle \overset{1.0}{\mapsto} \langle id, \texttt{a}, 11 \rangle$ would work on LSAs with any identifier.

This simple language is extended with some key ingredients to fit the goals of our framework. First of all, to allow interaction between different LSA-spaces, we introduce the concept of *remote pattern*, written $+P$, which is a pattern that will be matched with an LSA occurring in a neighbouring LSA-space (called a *remote LSA*): for example, $\langle id, \mathtt{a}, 10 \rangle \overset{1.0}{\longmapsto} +\langle id, \mathtt{a}, 11 \rangle$ removes the reagent LSA in a space (called the *local space*), and injects the product LSA into a neighbouring space, called the *remote space* (chosen probabilistically among matching neighbours). Note that when more remote patterns occur into an eco-law, they are all assumed to work on the same remote space, e.g., $P_1, +P_2, +P_3 \overset{r}{\longmapsto} P_4, +P_5$ works on a local space where P_1 occurs and a remote space where P_2, P_3 occur, and its effect is to replace P_1 with P_4 locally, and P_2, P_3 with P_5 remotely.

In order to allow eco-laws to apply to a wide range of LSAs, the argument of a pattern can also be a mathematical expression—including infix/prefix operators, e.g., $+, -, *, /, min$. For instance, $\langle id, \mathtt{a}, x \rangle \overset{1.0}{\longmapsto} \langle id, \mathtt{a}, x + 1 \rangle$ makes third argument of a matching LSA be increased by 1. Finally, among variables, some system variables can be used both in reagents and products to refer to contextual information provided by the infrastructure; they are prefixed with $\#$ and include $\#T$ which is bound to the time at which the eco-law fires, $\#D$ which is the topological distance between the local space and the remote space, and $\#O$ which is the orientation of the remote space with respect to the local space—e.g., an angle, a `north`/`south`/`west`/`east` indication, or any useful term like `in-front-of`, `on-rigth`, `in-the-same-room`.

4   A Crowd Evacuation Application

A public exposition is taking place in a museum composed of rooms, corridors, and exit doors. The surface of the exposition is covered by sensors, arranged in a grid, able to sense fire, detect the presence of people, interact with other sensors in their proximity as well as with PDAs that visitors carry with them.

When a fire breaks out, PDAs (by interaction with sensors) must show the direction towards an exit, along a safe path. The system has to be resilient to changes or unpredicted situations, in particular the safe path should: *(i - distance)*: lead to the nearest exit; *(ii - fire)*: stay away from fire; and *(iii - crowd)*: avoid overcrowded paths. These factors influence PDAs by means of the following LSAs:

– The exit gradient: each node contains an LSA with a numeric value indicating the distance from the nearest exit. These LSAs form a gradient field covering the whole expo [12,23]. Over an exit, the gradient value is 0, and it increments with the distance from the exit. When there are several exits or several paths towards an exit, the lowest distance value is kept for a node.
– The fire gradient: each node also contains an LSA indicating its distance from nearest fire. The LSA's value is 0 at the fire location and increases with the distance from it. It reaches a maximum value at a safe distance from a fire.
– The crowding gradient: sensors in the expo surface detect the presence of people and adjust the value in a local LSA indicating the crowding level of

the location. As with exit and fire, this LSA is diffused around, and its value increases with the distance from the crowded region.

- The attractiveness information: finally, each node contains an LSA indicating how desirable it is as a position in an escape route. Its value is based on the values of the previous three, and is used to choose which direction displays should point to.

4.1 Types of LSAs in the System

There are three forms of LSAs used in this scenario:

$$\langle source, type, max, ann \rangle, \ \langle grad, type, value, max, ann \rangle, \ \langle info, type, value, tstamp \rangle$$

A **source** LSA is used for gradient sources: *type* indicates the type of gradient (fire, exit, and so on); *max* is the maximum value the gradient can assume; and *ann* is the *annealing* factor [4]—its purpose will be described later, along with eco-laws. A **gradient** LSA is used for individual values in a gradient: *value* indicates the individual value; and the other parameters are like in the source LSAs. Finally, an **info** LSA is used for local values (e.g., not part of a gradient)—parameters are like in the source and gradient LSAs. The *tstamp* reflects the time of creation of the LSA.

4.2 Building the Fire, Exit and Crowding Gradients

The sources of the gradients are injected by sensors at appropriate locations, with the values $\langle source, exit, Me, Ae \rangle$ and $\langle source, fire, Mf, Af \rangle$. For the crowding information, we may assume that sensors are calibrated so as to locally inject an LSA indicating the level of crowding. The actual threshold in number of people will mainly depend on the sensor arrangement, and should be seen as a configuration issue. The crowding LSA looks like $\langle source, crowd, Mc, Ac \rangle$ and is periodically updated by the sensor.

As sources are established, gradients are built by the following set of eco-laws, applying to exit, fire and crowding gradients by parameterising argument *type* in the LSAs. First, we define Eco-law 1 that, given a source, initiates its gradient:

$$\langle source, T, M, A \rangle \xrightarrow{R_{init}} \langle source, T, M, A \rangle, \langle grad, T, 0, M, A \rangle \qquad (1)$$

When a node contains a gradient LSA, it spreads it to a neighbouring node with an increased value, according to Eco-law 2:

$$\langle grad, T, V, M, A \rangle \xrightarrow{R_s} \langle grad, T, V, M, A \rangle, +\langle grad, T, \min(V + \#D, M), M, A \rangle \qquad (2)$$

Due to use of system variable $\#D$, each node will carry a **grad** LSA indicating the topological distance from the source. When the spread values reach the maximum value M, the gradient becomes a plateau. Also note that the iterative application of this eco-law causes continuous diffusion of the LSA to all neighbouring nodes.

The spreading eco-law above may produce duplicate values in locations (due to multiple sources, multiple paths to a source, or even diffusion of multiple LSAs over time). Thus, Eco-law 3 retains only the minimum distance:

$$\langle \texttt{grad}, T, V, M, A \rangle, \langle \texttt{grad}, T, W, M, A \rangle \rightarrow \langle \texttt{grad}, T, \min(V, W), M, A \rangle \qquad (3)$$

Finally, we have to address the dynamism of the scenario where people move, fires extinguish, exits may be blocked, crowds form and dissolve. For instance, if a gradient source vanishes, the diffused values should increase (the distance to exit increases if the nearest exit is no longer available). However, with the above eco-laws this does not happen because Eco-law 3 always retain the minimum value. This is the purpose of the annealing parameter in the gradient LSAs: it defines the rate of Eco-law 4, which continuously tends to level up gradient values, encouraging the replacement of old values by more recent ones:

$$\langle \texttt{grad}, T, V, M, A \rangle \xmapsto{R_{ann}(A)} \langle \texttt{grad}, T, V{+}1, M, A \rangle \qquad (4)$$

The R_{ann} rate is directly proportional to A. When a fire is put out, for example, this eco-law will gradually raise the fire gradient to the point where it reaches the maximum, indicating no fire. Annealing may introduce a burden on the system, therefore high annealing values should only be used for gradients that have to change often or quickly.

4.3 Ranking Escape Paths: The Attractiveness Field

Based on exit distance, fire distance and crowding, a location can be ranked as more or less "attractive" to be part of an escape path. This is done via an *attractiveness* value automatically attached to each node by Eco-law 5:

$$\langle \texttt{grad}, \texttt{exit}, E, Me, Ae \rangle, \langle \texttt{grad}, \texttt{fire}, F, Mf, Af \rangle, \langle \texttt{info}, \texttt{crowd}, CR, TS \rangle \xmapsto{R_{att}}$$
$$\langle \texttt{grad}, \texttt{exit}, E, Me, Ae \rangle, \langle \texttt{grad}, \texttt{fire}, F, Mf, Af \rangle, \langle \texttt{info}, \texttt{crowd}, CR, TS \rangle, \qquad (5)$$
$$\langle \texttt{info}, \texttt{attr}, (Me - E)/(1 + (Mf - F) + k \times (Mc - C)), \#T \rangle$$

Coefficient k (tuned by simulation) is used to weight the effect of crowding on attractiveness. As gradients evolve, older attractiveness LSAs are replaced with newer ones (T is assumed positive):

$$\langle \texttt{info}, \texttt{attr}, A, TS \rangle, \langle \texttt{info}, \texttt{attr}, A2, TS{+}T \rangle \rightarrow \langle \texttt{info}, \texttt{attr}, A2, TS{+}T \rangle \qquad (6)$$

4.4 Choosing a Direction

Each location contains by default an LSA of the form $\langle \texttt{info}, \texttt{escape}, L, TS \rangle$, where L is the direction to be suggested by the PDA. In principle, the neighbour with the highest attractiveness should be chosen, but a more resilient solution is to tie the markovian rate of eco-laws to the attractiveness of neighbours, so that the highest probability is to point the best neighbour, with a possibility to point a less-than-optimal (but still attractive) neighbour:

$$\begin{aligned}&\langle\mathtt{info},\mathtt{escape},L\rangle,\langle\mathtt{info},\mathtt{attr},A,TS\rangle,+\langle\mathtt{info},\mathtt{attr},A+\Delta,TS2\rangle \xmapsto{R_{disp}(\Delta)}\\&\langle\mathtt{info},\mathtt{escape},\#O\rangle,\langle\mathtt{info},\mathtt{attr},A,TS\rangle,+\langle\mathtt{info},\mathtt{attr},A+\Delta,TS2\rangle\end{aligned} \qquad (7)$$

The rate is proportional to the difference in attractiveness between the node and its neighbour (Δ). The higher is Δ, the higher is the rate. Note that Δ is assumed to be positive value for otherwise the rate would be negative, hence $A+\Delta$ implies that Eco-law 7 only considers neighbours with a higher attractiveness, i.e., the PDA will not point *away* from the exit. Finally, once a direction is chosen, people move of discrete steps inside the environment. It is also defined a minimum possible distance between them, so to model the physical limit and the fact that two visitors can't be in the same place at the same time.

4.5   Resilient Behaviour

We adopt the definition of "resilient system" as one that is *tolerant to the harmful faults, which are most likely to affect the overall system service. [...] Fault tolerance is an acute issue in designing resilient pervasive systems"* [10]. The proposed architecture allows to express models that adapt dynamically to unexpected events (e.g., here we may have node failures, network isolation, exits suddenly unavailable, crowd formation, and so on) while maintaining their functionality. The nature of the eco-law language, akin to a rewriting system, is particularly suitable to express models by composing simple patterns that are known to exhibit resilient behaviour, thus not having to explicitly express aspects like error recovery or exceptional behaviour. One such pattern, for example, is the one building gradients. For a discussion more focused on patterns, see [21,7]. We will now discuss a few examples of possible problems and show how our example system reacts to them.

Single-node failure. Most behaviour in this architecture is based on nodes being able to apply eco-laws and host LSAs. Failure of a node clearly impacts this behaviour in some measure. A failing node (i.e., disappearing from the system) results in a physical location where no information for the displays is available, and where gradient information is not received or transmitted.

If the failing node is a generic one (i.e, no fire or exits at that node) and its disappearance does not cause network isolation, the impact is somewhat limited. Gradients will still be transmitted *around* the broken node, although values will change. Displays traversing a failing node will not update their direction; however, they will also never guide people towards a failing node, because it does not have an attractiveness value; they will rather steer people around the failing node. This means that functionality is preserved with decreased efficiency (i.e., longer paths).

If the failing node is a node containing a fire, the consequences are more severe: the fire will not be detected. PDAs will still never point the failing node on fire directly, but they will let people pass near it. One might argue that people will still see the fire; however, from the point of view of system behaviour, safety is reduced. On the other hand, it can be assumed that a serious fire would be detected by more sensors, and that all of them failing at the same time is unlikely.

If the failing node is an exit node, functionality gets a major hit, because the whole system will lack a fundamental information for display functioning (the exit gradient). If other exits are available, the system will still guide people towards them (again preserving functionality with reduced efficiency). However, if the failing node "hides" the only available exit, the system completely fails to provide useful information. This situation can be tackled by redundant source nodes near each exit.

Network isolation. When a group of failing nodes is the only connection between two parts of the network, this causes network isolation. There are a few sub-scenarios to consider. If the isolated section does not include exits, we find another limitation: the evacuation mechanism cannot work inside that section, as it is lacking fundamental information. If the section does however contain exits, it will act as an autonomous system in itself. Efficiency may be reduced but functionality will still be preserved, with an important caveat: this section of the network will ignore fires and alarms occurring in the rest of the network. This may or may not be an acceptable limitation depending on the scenario. For fires, it could probably be unacceptable. In a general way, however, we would consider that an isolated part of the network behaving as an autonomous system is an acceptable fallback. Again, redundancy is desirable for those nodes of the network that can cause isolation.

Exit unavailability. If an exit is suddenly unavailable in the expo, for instance it is broken and hence it does not open, a sensor could detect this and drop the source LSA. Because of the annealing mechanism, this causes the gradient value in each node to converge to the situation corresponding to the unavailability of that door; namely, a new path toward another exit emerges and affects the behaviour of all PDAs—people would simply go back to another exit as expected.

Crowd formation. While people follow their PDAs towards the nearest exit, it is possible that a jam forms in front of certain doors or corridors: this is one of the most perilous, unexpectable situations to experience in crowd evacuation. The application we set up is meant to emergently tackle these situations by means of the crowding gradient.

Assume two available corridors exist towards some exit, and from a given room people can choose to walk through any of them. If one becomes unavailable because a jam is forming the crowding gradient would emerge, reducing attractiveness along that corridor. Accordingly, it is expected that people behind the crowded area start walking back and head for the other corridor, and people in the room start choosing systematically the path free from crowd, even if they do not "see" the problem from the room.

5   Simulation

The proposed system is specified by a limited set of eco-laws, equipped with CTMC semantics. This specification can be used to feed the SAPERE infrastructure and be simulated in order to check the model dynamics in advance.

In this section we describe the prototype simulator we used, along with some simulations to validate the general correctness of the proposed system and to tune some parameter of the model before implementation and deployment.

5.1   The Simulator

In order to simulate the scenario described in Section 4, the main requirement for a simulation engine is to support a computational model built around a set of interacting and mobile nodes, which autonomously perform internal actions to aggregate/transform the information they hold. Such actions have the form of eco-laws that change system state following the CTMC model. Moreover, the number and position of nodes can change over time to model for instance new PDAs entering the system and moving around, or nodes breaking down.

Although many works manage to capture complex scenarios like the above one, we found existing approaches not completely fitting our framework. For instance, the agent-based model (ABM) [11] is a computational approach which provides the useful abstractions for modelling some of the mentioned concepts, *i.e.*, each node can be modelled as an agent that owns an internal behaviour and interacting capabilities. An ABM is simulated on top of platforms such as Repast or MASON [18] which provide the user with model specification and execution services. Unfortunately they do not support the CTMC model, which is a key element of the eco-laws model.

This is instead typically supported by simulators of chemical systems. However, the state-of-the-art in this context does not focus on highly mobile networks of chemical compartments [13]. A good deal of work has actually been moving towards multi-compartimentalisation: from the single, global solution idea of e.g. stochastic π-calculus [17], to mechanisms and constructs tackling the multi-compartment scenario of *Membrane computing* [16] and Bio-PEPA [5]. The mentioned languages and frameworks are not however conceived to address systems composed by a huge number of interacting compartments, and do not support dynamic networks, for the topology of the system is static.

For the above reasons, in the context of the SAPERE project we are developing an ad-hoc simulator with the purpose of executing chemical-like, multi-compartment scenarios resembling pervasive ecosystems, in which the position and number of nodes can dynamically vary with time as a results of adding, removing and moving nodes in the system. For the sake of brevity, we only sketch here its basic features.

The behaviour of each node is programmed according to the eco-laws coordination model explained in Section 3. Moreover the concept of *reaction* in the classical form of $A+B \xrightarrow{r} C+D$ is extended, still maintaining the CTMC semantics, into the form $c_1, \ldots, c_i \xrightarrow{f(k,c_1,\ldots,c_i,a_1,\ldots,a_j)} a_1, \ldots, a_j$ where reactants are a

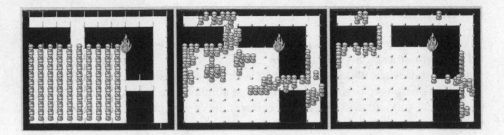

Fig. 1. A simulation run of the reference exposition: three snapshots

series of conditions over LSAs in local and remote LSA-spaces, and products are a series of changes to such LSA-spaces. The markovian rate is computed according to a kinetic function f. This model allows us to represent simple chemical reactions up to complex state transitions such as those modelled by eco-laws.

The simulator engine is written in Java and features: *(i)* an XML-based, low-level specification language to structure reactions, which can be used as sort of byte-code that higher-level specifications can be compiled to; *(ii)* an internal engine based on the "Next Reaction Method" (an optimised version of the well-known Gillespie's algorithm [9]); *(iii)* a set of configurable interfaces to show simulation results visually.

5.2   Simulation Setting

We here present simulations conducted over an exposition structured as shown in Figure 1, where three snapshots of a simulation run are reported: all the people in the room start moving towards one of the two exists (located at the ends of the corridor) because of the fire in the top-right corner of the room. Note in third snapshot that a person is walking in the middle of the corridor, for she was suggested to go to a farther exit because of the corridor jam at the bottom-right.

Rooms and corridors are covered by a grid of locations hosting sensors, one per meter in the room, one per two meters in the corridor: such locations are the infrastructure nodes where LSAs are reified. The maximum values for the gradients are set to: $Me = 30$, $Mf = 3$, $Mc = 20$. The PDA of each person is modelled as a mobile node, able to perceive the attractiveness gradient in the nearest sensor location: accordingly, the person moves in the suggested direction.

Figure 2 shows the gradients of exit, fire, crowding and attractiveness (one per column) corresponding to the simulation steps of Figure 1 (one step per row). At $t = 0$, gradients are level; with time, the gradient self-modify—it is easy to see the exits, fire and crowds in the respective gradients. The crowding gradient in the third column changes dynamically during simulation according to the movement of people. The last column shows the attractiveness gradient, computed from the other three gradients. Note how the second snapshot shows an attractiveness "hole" in the middle of the room and in the corridor due to crowding.

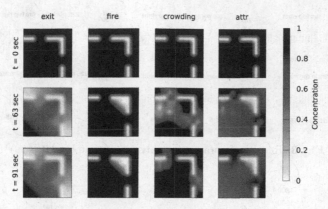

Fig. 2. Gradients for the snaphots in Figure 1. Concentration is the gradient value normalised to its maximum value.

5.3   Tuning Parameters

We chose to tune by simulation two parameters of the model: k (used in the eco-law computing attractiveness), and a multiplication factor p we applied to the crowding gradient to control its slope—namely, how far crowding should be perceived. Different simulations have been performed in order to find the set of parameters that minimises the time of exit of all the people in the room: the range of variation of k is $[0, 1]$ with a step of 0.1, while the range for p is $[0, 20]$ with step of 0.5. The smaller is k, the lower is the impact of the crowding in the computation of attractiveness, so that we expect the crowds not be avoided and the exit time to grow. The bigger is p, the higher is the crowding gradient slope, until it becomes a local information that PDAs can perceive only very close to it. Experiments have been done with 50, 100 and 225 persons in the room and with a fire in the corner of the room or in the middle, so as to experimentally observe the impact of parameters in the evacuation outcome of different scenarios. For each couple of parameters, 10 simulations have been run, considering average value and standard deviation of the resulting time.

In Figure 3 we show the results we obtained from the analysis of the parameters. Only dynamics for $k < 0.5$ are shown, because the time of exit is much higher elsewhere, and also because standard deviation becomes much higher since the system is much more chaotic.

The graphs show that with $k = 0$ (i.e., if the crowding gradient does *not* influence attractiveness) the performance of the system slows down as the number of persons increases, and the likelihood of crowd formation grows. For 50 and 100 persons not considering the crowding gradient does not have a big impact in the system dynamic as if very minimal congestion is formed, whereas it significantly increases the exit time with 225 persons. This result highlights the impact of considering crowds in resilience with respect to jams. For $k = 0.1$ we obtained the best performance for the system for both simulated fire positions. Results

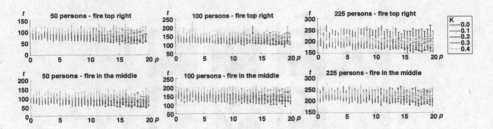

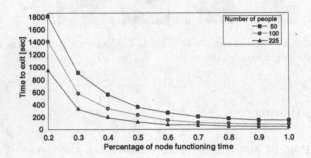

Fig. 3. Results of parameter analysis (t: time of exit [sec])

Fig. 4. Resilience to temporary node failures

finally showed that the parameter p does not seem to have relevant impact on the average exit time in this particular expo.

5.4 Resilience to Node Failures

Among the many simulations analysing resilience aspects, we focused on the case in which all nodes experience temporary failures, i.e., they work correctly only for a percentage of time. We expect the system to still be safe, i.e., be able to drive people out of the room in reasonable time, in spite of the changes in gradient values around the broken nodes. The results of simulation with 50, 100 and 225 persons are shown in Figure 4: the time to exit does not significantly increase even in the case of nodes being failing for 30%-40% of the time, and people are still able to eventually reach the exit even with 80% of node downtime.

6   Related Works

Chemical-oriented coordination. The proposed coordination model follows the tradition of approaches embedding coordination policies "within" a shared dataspace, following the pioneering works of approaches like TuCSoN [15] and LIME [14]. Our proposal extends these models to include bio-inspired ecological mechanisms, by fine-grained and well structured chemical-like reactions, which apply semantically. In particular, the coordination approach we propose in this paper originates from the chemical tuple space model in [22], with some notable differences: *(i)* here we provide a detail notational framework to flexibly express eco-laws that work on patterns of LSAs and affect their properties; *(ii)* the chemical concentration mechanisms proposed in [22] to exactly mimic chemistry is not mandatory here—though it can be achieved by a suitable design of rate expressions; *(iii)* the way we conceive the overall infrastructure and relationships between agents and their LSAs goes beyond the mere definition of the tuple-space model.

Beside tuple spaces, chemistry has been a source of inspiration for several works in (distributed) computing and coordination like in the Gamma language and its extensions such as HOCL [2] and CHAM [3]. The main features we inherit from this research thread include: *(i)* conferring a high-level, abstract, and nature-inspired character to the language used to program the distributed system behaviour; *(ii)* providing a reactive computational model very useful in autonomic contexts. While Gamma and it extensions (such as HOCL) were exploited in different application contexts [2], they originated with the goal of writing concurrent, general-purpose programming languages. Our approach instead aims at specifically tackling coordination infrastructures for pervasive systems, which calls for dictating specific mechanisms of diffusion, context- and spatial-awareness.

Situatedness and Context-Awareness. Considering the issues of situatedness and context-awareness, extensions or modifications to traditional approaches have been recently proposed to address adaptivity in pervasive environments. Similarly to our approach, in PLASTIC [1] service descriptions are coupled with dynamic annotations related to the current context and state of a service, to be used for enforcing adaptable forms of service discovery. However, our approach gets rid of traditional discovery services and enforces dynamic and adaptive service interaction via simple chemical reactions and a minimal middleware. In many proposals for pervasive computing environments and middleware infrastructures, the idea of "situatedness" has been promoted by the adoption of shared virtual spaces for services and components interactions. The pioneering system Gaia [19] introduces the concept of active spaces, that is active blackboard spaces acting as the means for service interactions. Later on, a number of Gaia extensions where proposed to enforce dynamic semantic pattern-matching for service composition and discovery [8] or access to contextual information [6]. Our model shares the idea of conceiving components as "living" and interacting in a shared spatial substrate (of tuple spaces) where they can automatically discover and interact with one another. Yet, our aim is broader, namely, to dynamically and systemically enforce situatedness, service interaction and data management with a simple language of chemical reactions.

7   Conclusions

This article discussed the SAPERE approach to modeling self-organising and self-adaptive ecosystems. We briefly discussed the approach and overviewed a fragment of the SAPERE model and eco-laws language. We gave an example of application of the approach to model crowd evacuation, and validated the approach using simulation.

The example we made shows that the SAPERE approach is well-suited to support generalised descriptions of self-organising behavioural patterns. Through parameterisation of eco-laws and LSAs, the same set of rules could be used for different gradients. On the one hand this simplifies modeling, on the other hand it supports scalability and expandability. This is especially important if one does not want to impose closed ecosystems. A further level of openness can be achieved by semantic matching (not discussed in this paper).

SAPERE models can achieve resilience without the need to explicitly model error recovery or exceptional behaviour. This is thanks to their self-adaptive nature and to the possibility to modeling by simple patterns known to be resilient.

This project has started recently, and these are the first analyses of concrete scenarios. Perspectives for the immediate future include performing a larger set of simulations, including larger expo environments, effects of crowd formation in steering, and so on; analysis, modeling and simulation of further scenarios, with different types of complexity; a further refinement of the LSA and eco-law syntax (more readable and user-friendly); and finally, more advanced validation techniques, like probabilistic symbolic model checking.

References

1. Autili, M., Di Benedetto, P., Inverardi, P.: Context-aware adaptive services: The PLASTIC approach. In: Chechik, M., Wirsing, M. (eds.) FASE 2009. LNCS, vol. 5503, pp. 124–139. Springer, Heidelberg (2009)
2. Banâtre, J.P., Priol, T.: Chemical programming of future service-oriented architectures. Journal of Software 4, 738–746 (2009)
3. Berry, G., Boudol, G.: The chemical abstract machine. In: POPL, pp. 81–94 (1990)
4. Casadei, M., Gardelli, L., Viroli, M.: Simulating emergent properties of coordination in Maude: the collective sort case. In: Proceedings of FOCLASA 2006. Electronic Notes in Theoretical Computer Science, vol. 175(2), pp. 59–80. Elsevier Science B.V., Amsterdam (2007)
5. Ciocchetta, F., Duguid, A., Guerriero, M.L.: A compartmental model of the cAMP/PKA/MAPK pathway in Bio-PEPA. CoRR abs/0911.4984 (2009)
6. Costa, P.D., Guizzardi, G., Almeida, J.P.A., Pires, L.F., van Sinderen, M.: Situations in conceptual modeling of context. In: EDOC 2006, p. 6. IEEE-CS, Los Alamitos (2006)
7. Fernandez-Marquez, J.L., Arcos, J.L., Di Marzo Serugendo, G., Viroli, M., Montagna, S.: Description and composition of bio-inspired design patterns: the gradient case. In: Proceedings of the 3rd Workshop on Bio-Inspired and Self-* Algorithms for Distributed Systems, ACM, Karlsruhe (2011)
8. Fok, C.L., Roman, G.C., Lu, C.: Enhanced coordination in sensor networks through flexible service provisioning. In: Field, J., Vasconcelos, V.T. (eds.) COORDINATION 2009. LNCS, vol. 5521, pp. 66–85. Springer, Heidelberg (2009)
9. Gibson, M.A., Bruck, J.: Efficient exact stochastic simulation of chemical systems with many species and many channels. The Journal of Physical Chemistry A 104(9), 1876–1889 (2000)

10. Iliasov, A., Laibinis, L., Romanovsky, A., Sere, K., Troubitsyna, E.: Towards rigorous engineering of resilient pervasive systems. In: Proceedings of Seventh European Dependable Computing Conference, IEEE Computer Society, Los Alamitos (2008)
11. Macal, C.M., North, M.J.: Tutorial on agent-based modelling and simulation. Journal of Simulation 4, 151–162 (2010)
12. Mamei, M., Zambonelli, F.: Programming pervasive and mobile computing applications: The tota approach. ACM Trans. Softw. Eng. Methodol. 18(4), 1–56 (2009)
13. Montagna, S., Viroli, M.: A framework for modelling and simulating networks of cells. In: Proceedings of the CS2Bio 2010 Workshop. ENTCS, vol. 268, pp. 115–129. Elsevier Science B.V, Amsterdam (2010)
14. Murphy, A.L., Picco, G.P., Roman, G.C.: Lime: A model and middleware supporting mobility of hosts and agents. ACM Trans. on Software Engineering and Methodology 15(3), 279–328 (2006)
15. Omicini, A., Zambonelli, F.: Coordination for Internet application development. In: Autonomous Agents and Multi-Agent Systems, vol. 2(3), pp. 251–269 (September 1999), special Issue: Coordination Mechanisms for Web Agents, http://springerlink.metapress.com/content/uk519681t1r38301/
16. Paun, G.: Membrane Computing: An Introduction. Springer-Verlag New York, Inc., Secaucus (2002)
17. Priami, C.: Stochastic pi-calculus. The Computer Journal 38(7), 578–589 (1995)
18. Railsback, S.F., Lytinen, S.L., Jackson, S.K.: Agent-based simulation platforms: Review and development recommendations. Simulation 82(9), 609–623 (2006)
19. Román, M., Hess, C.K., Cerqueira, R., Ranganathan, A., Campbell, R.H., Nahrstedt, K.: Gaia: a middleware platform for active spaces. Mobile Computing and Communications Review 6(4), 65–67 (2002)
20. Roy, P.V., Haridi, S., Reinefeld, A., Stefany, J.B.: Self management for large-scale distributed systems: An overview of the SELFMAN project. In: de Boer, F.S., Bonsangue, M.M., Graf, S., de Roever, W.-P. (eds.) FMCO 2007. LNCS, vol. 5382, pp. 153–178. Springer, Heidelberg (2008)
21. Tchao, A., Risoldi, M., Di Marzo Serugendo, G.: Modeling self-* systems using chemically-inspired composable patterns. In: Proceedings of the 5th IEEE International Conference on Self-Adaptive and Self-Organizing Systems. IEEE-CS, Los Alamitos (2011)
22. Viroli, M., Casadei, M.: Biochemical tuple spaces for self-organising coordination. In: Field, J., Vasconcelos, V.T. (eds.) COORDINATION 2009. LNCS, vol. 5521, pp. 143–162. Springer, Heidelberg (2009)
23. Viroli, M., Casadei, M., Montagna, S., Zambonelli, F.: Spatial coordination of pervasive services through chemical-inspired tuple spaces. ACM Transactions on Autonomous and Adaptive Systems 6(2), 14:1–14:24 (2011)
24. Viroli, M., Zambonelli, F.: A biochemical approach to adaptive service ecosystems. Information Sciences 180(10), 1876–1892 (2010)

Towards a Model-Driven Infrastructure
for Runtime Monitoring*

Antonia Bertolino[1], Antonello Calabrò[1], and Francesca Lonetti[1],
Antinisca Di Marco[2], and Antonino Sabetta[3],**

[1] Istituto di Scienza e Tecnologie dell'Informazione "A.Faedo", CNR, Pisa, Italy
{firstname.secondname}@isti.cnr.it
[2] Department of Computer Science, University of L'Aquila, Coppito (AQ), Italy
antinisca.dimarco@univaq.it
[3] SAP Research, Sophia Antipolis, France
antonino.sabetta@sap.com

Abstract. In modern pervasive dynamic and eternal systems, software
must be able to self-organize its structure and self-adapt its behavior
to enhance its resilience and provide the desired quality of service. In
this high-dynamic and unpredictable scenario, flexible and reconfigurable
monitoring infrastructures become key instruments to verify at runtime
functional and non-functional properties. In this paper, we propose a
property-driven approach to runtime monitoring that is based on a comprehensive Property Meta-Model (PMM) and on a generic configurable
monitoring infrastructure. PMM supports the definition of quantitative
and qualitative properties in a machine-processable way making it possible to configure the monitors dynamically. Examples of implementation
and applications of the proposed model-driven monitoring infrastructure
are excerpted from the ongoing CONNECT European Project.

1 Introduction

Nowadays software systems are increasingly pervasive and dynamic, and their
(lack of) quality may have a deep impact on businesses and people's lives. At the
same time, as we entrust more and more responsibilities to distributed software
systems, the need arises to augment them with powerful oversight and management functions in order to allow continuous and flexible monitoring of their
behavior.

In this perspective, runtime monitoring is the key technological enabler both
for quality assurance and for prolonging software lifecycle after deployment,
by supporting runtime verification and online adaptation. Indeed, it provides
a mean for evaluating and enhancing the resilience of dynamic and evolvable
systems allowing the system to recover from abnormal situations or to prevent
them by taking proactive actions.

* This work is partially supported by the EU-funded CONNECT project (FP7–231167)
and EU-funded VISION ERC project (ERC-240555).
** Antonino Sabetta contributed to this paper while he was a researcher at CNR-ISTI,
Pisa.

E.A. Troubitsyna (Ed.): SERENE 2011, LNCS 6968, pp. 130–144, 2011.

Putting in place a working monitoring process involves several activities [17], including the collection of raw observation data, the interpretation of such raw information to recognize higher-level events that are relevant at the business level, and the effective presentation of the results of the monitoring. In real systems, the volume of collected raw data can be overwhelming. Moreover, such data need to be filtered and aggregated to detect deviations from expected behavior and to derive measurements of interesting non-functional properties of the system.

There is a gap to be filled when a high-level, business-relevant property is to be turned into a concrete setup of the monitoring infrastructure. In order to be able to evaluate and keep under control any such property, it is necessary to instruct the monitor about what raw data to collect and how to infer whether or not a desired property is fulfilled. Unfortunately, not only this task is time-consuming, but it also requires a substantial human effort and specialized expertise in order to convert the high-level description of the property to observe into lower-level monitor configuration directives. Consequently, the outcome of such effort is very hard to generalize and to reuse, and, as a matter of fact, the resulting monitor configuration typically is only relevant in the specific situation at hand. This process needs to be iterated each time a different architecture needs to be monitored, or when the properties to be monitored change.

To address the shortcomings of this process, in this paper we propose a model-driven approach [26] to monitoring that is based on two key elements: a generic monitoring infrastructure that offers the greatest flexibility and adaptability; and a coherent set of domain-specific languages, expressed as meta-models, to define models of properties that enable us to exploit the support to automation offered by model-driven engineering techniques. In the proposed approach, the properties to be monitored (either qualitative or quantitative) are specified according to a meta-model. Using this approach, and leveraging an underlying generic monitoring infrastructure, we can thus separate the problem of defining properties and metrics of interest, from the problem of converting these specifications into a concrete monitoring setup, which is done automatically in our approach.

The contribution of our proposal stays in: i) the effort (ambition) to offer a comprehensive meta-model for system properties that spans over dependability, performance, security and trust attributes by going beyond the state of the art since existing meta-models generally address only a subset of the above properties or do not support transformational approaches; ii) the inter-connection between the above meta-model and a modular event-based monitoring infrastructure.

The defined Property Meta-Model (PMM) is implemented as an eCore model and is provided with an editor. The ultimate vision we want to achieve is that a software developer using PMM can either retrieve from the editor repository the pre-built specification of a simple or complex metrics/property or, if not present, can build new properties and metrics to be monitored, using the concepts in the meta-model. Also a non-expert domain user can benefit from the pre-built models. By transformation these property models will be translated into the input format for the monitor in order to be observed and verified at runtime.

Our research on monitoring originated in the context of the FP7 "ICT forever yours" European Project CONNECT[1]. The CONNECT world envisions dynamic environments populated by technological islands which are referred to as the Networked Systems (NSs), and by the components of the CONNECT architecture, called the CONNECT Enablers. The ambitious goal of the project is to have eternally functioning systems within a dynamically evolving context, which is to be achieved by synthesizing *on-the-fly* the CONNECTors through which NSs communicate. The resulting emergent CONNECTors then compose and further adapt the interaction protocols run by the CONNECTed System. Evidently, such a dynamic context strongly relies on functional and non-functional behavior monitoring.

In the remaining part of this paper, we describe: the Property Meta-Model we defined to specify properties (Section 2); the GLIMPSE monitoring infrastructure that implements the Generic Monitoring Framework (Section 3); and finally the Monitor Configuration steps that from the PMM property models generate the code used to configure GLIMPSE to monitor the relevant properties (Section 4). An application example (Section 5), related work (Section 6), and conclusions (Section 7) complete the paper.

2   Property Meta-model

In this section, we give an overview of the Property Meta-Model (PMM) we defined for specifying observable properties of the system. Figure 1 sketches the key concepts of this meta-model that are: *Property*, *MetricsTemplate*, *Metrics*, *EventSet*, *EventType*, *ApplicationDomain*, and how they relate each other.

The Property Meta-Model describes a property that can be *ABSTRACT*, *DESCRIPTIVE*, or *PRESCRIPTIVE*. An *ABSTRACT* property indicates a generic property that doesn't specify a required or guaranteed value for an observable or measurable feature of a system. A *DESCRIPTIVE* property represents a guaranteed/owned property of the system while a *PRESCRIPTIVE* one indicates a system requirement. In both cases, the property is defined taking into account a relational operator with a specified value. A property can be qualitative or quantitative: the former is about events that are observed and cannot generally be measured, referring to the behavioral description of the system (e.g., deadlock freeness or liveness); the latter deals with quantifiable/measurable observations of the system and it has an associated *Metrics*. The *QuantitativeProperty* can have a *Workload* and an *IntervalTime*. The *Workload* can be open or close. To clarify the above concepts we report below two Property examples:

Property1: The system S in average responds in 3 ms in executing the e_1 operation with a workload of 10 e_2 concurrent operations.

---

[1] http://connect-forever.eu

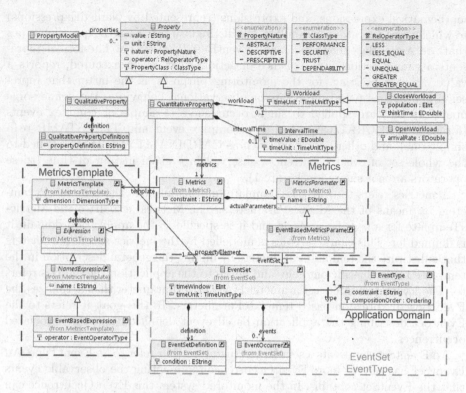

Fig. 1. Key concepts of the Meta-Model for processable properties

Property2: The system S in average must respond in 3 ms in executing the e_1 operation with a workload of 10 e_2 concurrent operations.

The former is a *DESCRIPTIVE* property while the latter is a *PRESCRIPTIVE* one because it specifies a required time response with a value of 3 ms in executing the e_1 operation. Both are quantitative properties having a PERFORMANCE class because they refer to a measurable performance dimension (time).

In the proposed meta-model, we make a distinction between a generic metrics (represented by a MetricsTemplate) and the concrete metrics (i.e., the Metrics concept) instantiated to a specific application domain, represented by the domain of the software system the PMM is used for. The MetricsTemplate represents the way a metrics can be specified. Hence a metrics (for example the response time) can refer to one or more (hopefully equivalent) specifications represented by different MetricsTemplates. The response time of an operation (E), for example, can be expressed as the duration of that operation or the difference between the timestamps of the ending (E1) and starting (E2) actions of that operation. E, E1 and E2 are templateParameters (not showed in Figure 1). A MetricsTemplate contains the definition of mathematical operators, nested into complex mathematical expressions, and expressions (represented by NamedExpressions) to which these operators are applied. We distinguish two NamedExpression types that are ActionBasedExpression and EventBasedExpression. The former represents a simple action or a sequence of actions that, when executed, reports a value. More interesting, for the monitoring purposes, is the latter that represents expressions based on events or observational behaviors. We define some operators that are applied to single occurrences of simple or complex events (for example DURATION refers to a complex event and TIMESTAMP to a simple one), and other operators (such as CARDINALITY) that are applied to the whole set of event occurrences observed in a given instant of time (these operators are not showed in Figure 1).

A metrics refers to a MetricsTemplate and instantiates the templateParameters by means of the MetricsParameters. The Metrics actualizes the MetricsTemplate for a specific scenario and it is specific to the application domain it is defined for. This characteristics is modelled by the metrics actual parameters that substitute the templateParameters by linking the general description in the template to the specific ontology and hence to the application the metrics refers to. The EventBasedMetricsParameter is a MetricsParameter that actualizes the EventBasedExpression based templateParameters. To this goal, it refers to the EventSet describing the application based event definition and the associated occurrences.

An EventSet represents a set of event instances that refer to an EventType. An EventSet has zero or more EventOccurrence representing the observable events that the EventSet contains. In the monitored system this EventOccurrence can be generated at runtime by the monitoring infrastructure when the probes observe the event of the EventType the EventSet refers to.

The EventType models an observable system behavior that can be a primitive/simple event or operation representing the lowest observable system activity or a composite/complex event that is a combination of primitive and other composite events. An EventType has a specification identifying the type or class of the observable events. Such EventTypeSpecification belongs to the ontology of a specific application domain. In the current version of the meta-model this specification is simply defined by means of a label or string. However, in the future we plan to provide a more formal specification for complex event definition, according to an existing or a new defined event specification language. In [11,25] some examples of complex event specification languages are presented.

The devised meta-model has been generated as an eCore model into the Eclipse Modeling Framework(EMF) [12]. In particular, we define the meta-model partitioned in the following eCore models: *Core.ecore* representing a generic named element, *EventType.ecore* and *EventSet.ecore* modeling the event and the eventSet respectively, *Metrics.ecore* and *MetricsTemplate.ecore* for specifying the metrics and metricsTemplate concepts and finally the *Property.ecore* representing the Property meta-model.

From the above defined eCore models, by means of the EMF facilities, an editor has been obtained as an Eclipse Plugin. This editor contains the information of the defined eCore models and allows to create new model instances of the Property, Metrics, MetricsTemplate, EventType and EventSet meta-models.

The presented Property Meta-Model has been used for defining properties relevant for the CONNECT project[2]. It proved to be complete for specifying all properties of interest.

The models conforming to such meta-model can be used to drive the instrumentation of the CONNECT monitoring Enabler that generates suitable probes to monitor useful properties on the CONNECTors.

3   GLIMPSE Architecture

In large and distributed systems, huge amount of events are generated and from their combination and filtering it is possible to timely detect unexpected behaviors of the systems or predict failures for enhancing the system resilience. GLIMPSE[3], is a flexible monitoring infrastructure, developed with the goal of decoupling the event specification from the analysis mechanism. GLIMPSE was initially proposed in the context of the CONNECT project, where it is used to support behavioral learning, performance and reliability assessment, security, and trust management. A prototype of the GLIMPSE architecture[4] has been developed and is being used in CONNECT. However, the infrastructure is totally generic and can be easily applied to different contexts. The architecture of GLIMPSE (shown in figure 2) is composed of five main components that implement the main five core functions identified into a generic monitoring infrastructure [17], namely: Data collection, Local interpretation, Data transmission, Aggregation, Reporting.

Probes (Collector/Data Suppliers). Probes intercept primitive events when they occur in the software and send them to the GLIMPSE Monitoring Bus. Probes are usually realized by injecting code into an existing software or by using proxies. In addition, they may be configured to use a primitive event filter in order to reduce the amount of generated raw data.

Monitoring Bus. The Monitoring Bus is the communication backbone that all information (events, questions, answers) is sent on: Probes, Consumers, Complex Event Processor and by all the services querying information to GLIMPSE. We adopt a publish-subscribe paradigm devoting the communication handling to the Manager component.

In the current GLIMPSE implementation, the system backbone is implemented by means of ServiceMix4 [5], an open source Enterprise Service Bus, used to combine advantages of event-driven architecture and service-oriented architecture functionality. We chose ServiceMix4 because it offers a Message Oriented Bus and is able to run an open source message broker like ActiveMQ [1].

---

[2] A first release of the CONNECT Property Meta-Model is available at
   http://labse.isti.cnr.it/tools/cpmm
[3] Generic fLexIble Monitoring based on a Publish-Subscribe infrastructurE
[4] Available at http://labse.isti.cnr.it/tools/glimpse

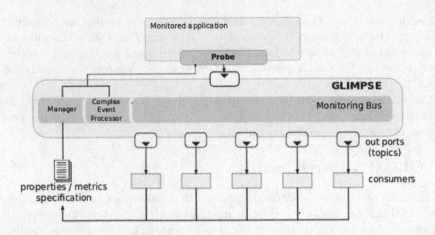

Fig. 2. Glimpse architecture

Complex Event Processor. The Complex Event Processor (CEP) is the rule engine which analyzes the primitive events, generated from the probes, to infer complex events matching the consumer requests. There are several rule engines that can be used for this task (like Drools Fusion [2], RuleML [4]).

In the current GLIMPSE implementation, we adopt the Drools Fusion rule language [2] that is open source and can be fully embedded in the realized Java architecture. Note that, the proposed flexible and modular architecture allows for easily replacing this specific rule language with another one.

Consumer. It may be a learning engine, a dependability analyzer or a simple customer that requests some information to be monitored. It sends a request to the Manager using the Monitoring Bus and waits for the evaluation results on a dedicated answer channel provided by the Manager.

Manager. The Manager component is the orchestrator of the GLIMPSE architecture. It manages the communications among the GLIMPSE components. Specifically, the Manager fetches requests coming from Consumers, analyzes them and instructs the Probes. Then, it instructs the CEP Evaluator, creates and notifies to the Consumer a dedicated channel on which it will provide results produced by the CEP Evaluator.

4   Property-Driven Monitoring Configuration

We have described in the previous sections the Property Modeling sub-process and the Generic Monitoring Infrastructure we intend to use. Here we briefly explain how these two sub-processes are combined, i.e., we describe the automatic "Monitors Configuration" we propose. Precisely, the editor provided along with PMM allows the software developer to specify a new property as a model that is

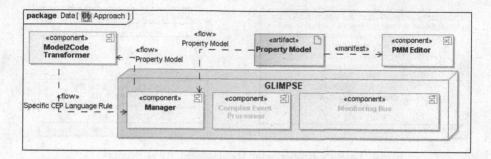

Fig. 3. Property-Driven GLIMPSE Configuration

conforming to the PMM meta-model. If this model represents a property to be monitored, it can be used to instruct the GLIMPSE infrastructure, according to the approach sketched in Figure 3. As shown, the GLIMPSE manager component takes in input such property model and activates an external component (named in the figure modelToCode Transformer), which performs the code generation according to the specific complex event processing language that is embedded into GLIMPSE. The output of this transformation is represented by a specific rule that is processed by the complex event processor component of GLIMPSE. For the sake of precision, at the time of writing the modelToCode Transformer component is still under development and the current running implementation of GLIMPSE infrastructure uses the Drools Fusion complex event processor [2]. Following the depicted property-driven approach, we are developing an automated modelToCode Transformer component (using Acceleo code generator tool) according to the Drools Fusion rule specification language and we are refining the meta-model in order to perform this. Indeed, the advantage of adopting a model-driven approach is that it allows the monitor to use any complex event processing engine as long as a modelToCode Transformer transforms the property model into the rule specification language of that processing engine. As an application example, in the next section we show in detail a property model, specified using the PMM meta-model, and the corresponding Drools Fusion rule.

5   Application Example

In this section we give an application example of our approach. Specifically, we first present the Terrorist Alert Scenario which is one of the demonstration examples chosen in CONNECT. Then, we show how to model using the PMM a latency property required in the system for that scenario, and finally how to express this property by means of the rule specification language used by Drools Fusion. Modeling of this latency property represents a simple example for giving an idea of the proposed approach. For the Terrorist Alert Scenario, we modeled, using the PMM, also dependability and security properties that we do no present here for space limitation reasons and we refer to [9] for their description.

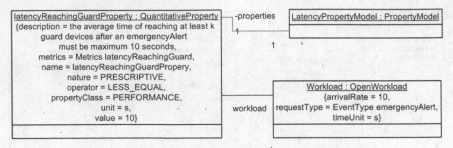

Fig. 4. Latency Property for the Terrorist Alert Scenario

Terrorist Alert scenario. The CONNECT Terrorist Alert scenario [10], depicts the critical situation that during the show the stadium control center spots one suspect terrorist moving around. The alarm is immediately sent to the Policemen, equipped with ad hoc handheld devices which are connected to the Police control center to receive commands and documents, for example a picture of a suspect terrorist. Unfortunately, the suspect is put on alert from the police movements and tries to escape, evacuating the stadium. The policeman that sees the suspect running away can dynamically seek assistance to capture him from civilians serving as private security guards in the zone of interest. To get help in following the moves of the escaping terrorist and capturing him, the policeman sends to the civilian guards an alert message in which a picture of the suspect is distributed. On their side, to perform their service, the guards that control a zone are CONNECTed in groups and are equipped with smart radio transmitters. The guards control center sends an EmergencyAlert message to all guards of the patrolling groups; the message reports the alert details. On correct receipt of the alert, each guard's device automatically sends an ack to the control center.

Latency Property for the Terrorist Alert Scenario. We show how to model the following required latency property: *average time needed by the CONNECTed system to reach k% guard devices must be at most equal to 10 seconds when in the system there are 10 alerts.* For "time needed by the CONNECTed system to reach a set percentage of guard devices" we mean the average latency experienced in the system from the incoming EmergencyAlert message to the reception of a percentage of eAck coming back from the reached guards' devices. The model for this PRESCRIPTIVE property is shown in Figure 4. This is a *PERFORMANCE* property requiring that the associated metrics is *LESS_EQUAL* to *10 s* when the system has a workload of *10* alerts. This metrics called *LatencyReaching-Guard* (omitted due to space limitations) is an instance of the Average Latency MetricsTemplate that is presented in Figure 5. We recall that the template is generic and the same model can be used in other scenarios. The average latency represents a TIME measure defined as average of the differences of the timestamps of two related generic event instances (x and y in the model), respectively as the latest event occurrence and the former one. Finally, the template exposes two templateParameters: e_1 bound to y, and e_2 bound to x. A Metrics, whose definition is an instance of the MetricsTemplate, concretises the template for a specific scenario. This is reflected in the metrics' actual parameters

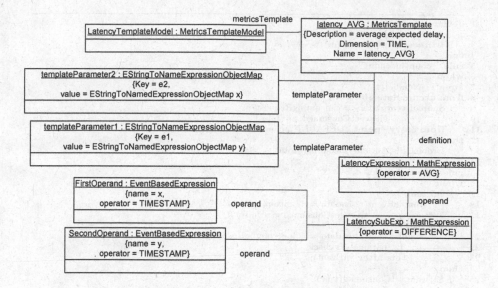

Fig. 5. Average Latency Metrics Template

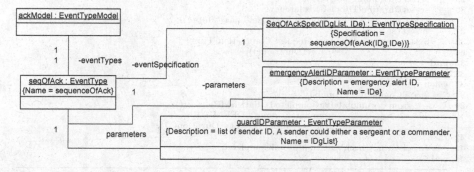

Fig. 6. Sequence of Ack for an Alert

which substitute the template parameters by linking the general description in the template to the specific ontology and hence to the application the metrics refers to. The *LatencyReachingGuard* metrics actualizes the corresponding Average Latency MetricsTemplate by linking to the templateParameters e_1 and e_2, two EventSets (e_1 and e_2 respectively), and specifying that the two event sets must satisfy a metrics constraint establishing that the two event sets must be related to each other. Referring to the Terrorist Alert scenario, the e_1 Event Set refers to the emergencyAlert Event Type that is a simple event definition since it corresponds to a message directly observable from the system. The e_2 EventSet instead refers to SeqOfAck EventType that is a sequence of eAck observed in the system. To be of interest of the *LatencyReachingGuard* metrics, the occurrences of SeqOfAck must contain at least k eAck occurrences related to each other, that means they refer to the same emergencyAlert message. The SeqOfAck Event-

```
1  declare SimpleEvent
2      @role(event)
3      @timestamp(timestamp)
4  end
5  rule "countIncoming"
6  when
7      $total : Number()
8  from accumulate($bEvent:
9      SimpleEvent(data == "incomingRequest",
10                 this.getConsumed == false)
11     from entry−point "DEFAULT", count ($bEvent))
12 then
13     DroolsUtils.ActuallyIncoming($total);
14 end
15 rule "checkCompleted"
16 when
17     $aEvent:
18         SimpleEvent(this.data == "incomingRequest",
19                     this.getConsumed == false);
20     $bEvent:
21         SimpleEvent(this.data == "outcomingResponse",
22             [... parameters check ...],
23             this after $aEvent);
24 then
25     $aEvent.setConsumed(true);
26     $bEvent.setConsumed(true);
27     SatisfiedRequest sr = new SatisfiedRequest();
28     sr.setIncoming($aEvent);
29     sr.setOutcoming($bEvent);
30     sr.setDuration(DroolsUtils.latency(
31        $aEvent.getTimestamp(),$bEvent.getTimestamp()));
32     insert (sr);
33     System.out.println("Last Request Completion Time: "
34                     + sr.getDuration());
35 end
36 rule "requestCompletedInTime"
37 when
38     $totalCompleted : Number()
39         from accumulate($aCompleted : SatisfiedRequest()
40                 from entry−point "DEFAULT",
41                 average($aCompleted.getDuration()))
42 then
43     DroolsUtils.CheckViolation($totalCompleted);
44 end
```

Listing 1.1. Drools latency rule example

Type, shown in Figure 6, is a complex EventType defined as a sequence of eAck simple EventTypes. It has two parameters: the emergencyAlert ID (namely IDe) the sequence refers to, and the list of guards messages acknowledging the alert (namely, IDgList). We recall that the occurrences of events are generated by the monitors at run-time when the system is running, once the probes observe the event of the EventType the EventSet refers to.

Drools rule specification for the Latency Property. The Listing 1.1 shows a fragment of the Drools rules used to monitor the Latency property for the Terrorist Alert Scenario depicted in Figure 4. Each event flowing on the GLIMPSE Monitoring Bus is an abstraction of a state transition of an LTS machine. In our LTS machine, the e_1 operation starts with the event incomingRequest and finishes with the outcomingResponse event. The first rule in Listing 1.1, `CountIncoming`, counts the incoming requests (lines 5-14). The `checkCompleted` rule (lines 15-35), analyzes the `incomingRequest` event and the `outcomingResponse` sent by the same client, using the fields: connectorInstanceID, connectorInstanceExecution, and creates a subset of accomplished requests (SatisfiedRequest) saving the computation execution time in the field Duration. Finally, the rule `requestCompletedInTime` (lines 36-44), evaluates the average time spent for completing requests.

6 Related Work

The approach proposed in this paper is strictly related to two areas of research: a) the specification of meta-models for defining software metrics and non-functional properties, and b) the design of runtime monitoring systems. This section presents a brief selection of relevant works from both these areas.

The main concept underlying our proposal, i.e., specifying metrics as instances of a metrics specification meta-model, is common to the work of Monperrus et al. [19], in which a generative model driven definition of software metrics is proposed. This work concerns the definition of a *domain-independent metrics meta-model*, allowing modelers to automatically add measurement capabilities to a domain specific modeling language used during the different phases of a model-driven development process (such as architectural design, requirements or implementation). Taking inspiration from the work of Monperrus et al., PMM separates the property and more specifically metrics definition from the application domain. Instead, differently from [19], PMM addresses specifically *non-functional property and metrics*. Indeed, it introduces additional concepts concerning the qualitative and quantitative properties definition, the events modeling and the distinction between a generic metrics (represented by a MetricsTemplate) and the concrete metrics (i.e., the Metrics concept) instantiated to a specific application domain.

Several works addressed meta-modeling focussing on domain-specific metrics or dependability properties [22,14]. For different reasons, these approaches propose partially what PMM proposes as a whole. PMM allows for the specification of different types of properties, such as, among the others, performance, security, dependability and trust properties and of relative metrics.

A more general Quality of Service Modeling Language (QML) is proposed in [13] for describing QoS specifications for software components in distributed object systems. It is an extension of UML, allowing a fine grained specification level of attributes and operations and a dynamic and runtime check of QoS components requirements and dependencies. Again, the UML MARTE profile [21] provides a common way for modeling hardware and software aspects of a real time embedded system. It provides facilities to annotate models with information required to perform quantitative predictions and performance analysis. A model-driven performance measurement and assessment approach is presented in [7]. It provides a meta-model for the specification of performance metrics and the possibility of specifying measurement points in the model. It allows for the automatic instrumentation and software code generation with integrated code for performance data collection, storage and metrics computation. An evaluation of the proposed approach is provided with an implementation of a UML profile and transformations from the profile to Java. Differently from the previous approaches, our proposal focuses on a more general and complete framework, that allows not only for specifying performance measures but enables the specification of qualitative and quantitative properties into a machine-processable language.

A promising research direction, addressing QoS modeling, focuses on ontologies that allow for the definition of QoS with rich semantic information. In particular, [16] presents a semantic QoS model addressing the main elements of dynamic service environments (networks, devices, application services and endusers). It makes use of Web Service Quality Model (WSQM) [20] standard to define QoS at the service level, and comprehends four ontologies specifying respectively: the core QoS concepts, the environment and underlying network and hardware infrastructure QoS properties, the application server and user-level QoS properties. As the authors claim in [16], their model concentrates on QoS knowledge representation rather than on a language to specify QoS. In this way, they provide separate and reusable ontologies and any appropriate QoS specification language can be used on top of it. Differently from this approach, PMM allows the specification of non-functional properties.

Concerning monitoring systems, the literature is rich of proposals of frameworks, languages, and architectures [23,15]. In particular, [23] presents an extended event-based middleware with complex event processing capabilities on distributed systems. Similar to GLIMPSE this work adopts a publish/subscribe infrastructure. Another monitoring architecture for distributed systems management is presented in [15]. This architecture employs a hierarchical and layered event filtering approach, specifically targeted at improving scalability and performance for large-scale distributed systems, minimizing the monitoring intrusiveness.

Defining expressive complex event specification languages has been an active research topic for years [25,8,11]. Among these languages, GEM [25] is a generalized and interpreted event monitoring language. It is rule-based (similar to other event-condition-action approaches) and provides a detection algorithm that can cope with communication delay. Snoop [8] follows an event-condition-action approach supporting temporal and composite events specification but it is especially developed for active databases. A more recent formally defined specification language is TESLA [11] that has a simple syntax and a semantics based on a first order temporal logic. The main focus of these works is the definition of a complex-event specification language, whereas our framework provides a more high-level and more specialized meta-model to define monitoring goals (functional properties and metrics definitions), which are then automatically translated into complex-event specifications.

Other monitoring frameworks exist that address the monitoring of performance, in the context of system management [6,18,3]. Among them, Nagios [6] offers a monitoring infrastructure to support the management of IT systems spanning network, OS, applications; Ganglia [18] is especially dedicated for high-performance computing and is used in large clusters, focusing on scalability through a layered architecture whereas the Java Enterprise System Monitoring Framework [3] deals with web-based and service-driven networks solutions.

Finally, the work in [24] has several similarities with our approach, concerning the conceptual modeling of non-functional properties. However it is more specifically focused on measurement refinement, whereas our work targets a more comprehensive scope for modeling and transformation. In future work we plan to look closely at this model to possibly incorporate some of its refinements.

7   Conclusions

We proposed a model-driven infrastructure for runtime monitoring. The monitoring configuration is automatically executed by parsing the models of the properties of interest. To allow for automatization, such models must conform to a suitable meta-model. In this paper we presented: *i)* a Property Meta-Model we devised to express properties and metrics. It allows for the definition of prescriptive/descriptive and qualitative/quantitative properties, and the metrics needed to quantify them. *ii)* GLIMPSE, which is an implementation of a generic monitoring infrastructure; and, *iii)* the model-driven monitor configuration, combining PMM and GLIMPSE. As proof of concept, we finally showed the application of the model-driven infrastructure for runtime monitoring to a CONNECT application scenario.

There are various directions for future work. First, as outlined in Section 4, we plan to implement the ModelToCode transformations to automatically derive the Drools rules needed to configure GLIMPSE. For what concerns PMM, some meta-model parts need to be refined. Among others, we need to refine: *i)* the EventTypeSpecification meta-class by introducing an event-based language that enables the specification of complex EventType; and *ii)* the QualitativeProperty-Definition meta-class, by defining a suitable language (meta-model) allowing for the specification of complex properties. Finally, so far PMM only supports the specification of the transition (or action)-based properties. However, state-based properties could be relevant in some cases. We plan to extend PMM in order to support also the modeling of state-based properties.

Moreover, in the future we want to address the reliability and performance issues of the proposed monitoring framework when a lot of data are generated, providing a comparison of GLIMPSE with similar existing approaches.

References

1. ActiveMQ: A complete message broker, http://activemq.apache.org
2. Drools Fusion: Complex Event Processor,
   http://www.jboss.org/drools/drools-fusion.html

3. Java Enterprise System Monitoring Framework,
   http://download.oracle.com/docs/cd/E19462-01/819-4669/geleg/index.html
4. Ruleml: The rule markup initiative, http://ruleml.org
5. ServiceMix: an open source ESB, http://servicemix.apache.org/home.html
6. Barth, W.: Nagios. System and Network Monitoring. No Starch Press, u.s (2006)
7. Bošković, M., Hasselbring, W.: Model Driven Performance Measurement and Assessment with MoDePeMART. In: Schürr, A., Selic, B. (eds.) MODELS 2009. LNCS, vol. 5795, pp. 62–76. Springer, Heidelberg (2009)
8. Chakravarthy, S., Mishra, D.: Snoop: An expressive event specification language for active databases. Data & Knowledge Engineering 14(1), 1–26 (1994)
9. CONNECT Consortium. Deliverable 5.2: Design of Approaches for dependability and initial prototypes (2011), http://connect-forever.eu/
10. CONNECT Consortium. Deliverable 6.1: Experiment scenarios, prototypes and report (2011), http://connect-forever.eu/
11. Cugola, G., Margara, A.: TESLA: a formally defined event specification language. In: Proceedings of DEBS, pp. 50–61 (2010)
12. Eclipse Platform, Eclipse Modeling Project, http://www.eclipse.org/modeling/
13. Frolund, S., Koistinen, J.: Quality-of-Service Specification in Distributed Object Systems. Distributed Systems Engineering Journal 5, 179–202 (1998)
14. Huhn, M., Zechner, A.: Analysing dependability case arguments using quality models. In: Buth, B., Rabe, G., Seyfarth, T. (eds.) SAFECOMP 2009. LNCS, vol. 5775, pp. 118–131. Springer, Heidelberg (2009)
15. Hussein, E., Abdel-wahab, H., Maly, K.: HiFi: A New Monitoring Architecture for Distributed Systems Management. In: Proceedings of ICDCS, pp. 171–178 (1999)
16. Mabrouk, N., Georgantas, N., Issarny, V.: A semantic end-to-end QoS model for dynamic service oriented environments. In: Proceedings of PESOS, pp. 34–41 (2009)
17. Masoud, M.S., Sloman, M.: Monitoring distributed systems. In: Network and Distributed Systems Management, pp. 303–347 (1994)
18. Massie, M., Chun, B., Culler, D.: The ganglia distributed monitoring system: design, implementation, and experience. Parallel Computing 30(7), 817–840 (2004)
19. Monperrus, M., Jézéquel, J., Baudry, B., Champeau, J., Hoeltzener, B.: Model-driven generative development of measurement software. In: Software and Systems Modeling, SoSyM (2010)
20. OASIS: Quality Model for Web Services (WSQM) (September 2005)
21. OMG: UML Profile for Modeling and Analysis of Real-Time and Embedded systems (MARTE), http://www.omg.org/omgmarte/Specification.htm/
22. Pataricza, A., Györ, F.: Towards unified dependability modeling and analysis. In: Proceedings of ARCS Workshops, pp. 113–122 (2004)
23. Pietzuch, P.R., Shand, B., Bacon, J.: Composite event detection as a generic middleware extension. IEEE Network 18(1), 44–55 (2004)
24. Röttger, S., Zschaler, S.: Tool Support for Refinement of Non-functional Specifications. Software and Systems Modeling 6(2), 185–204 (2007)
25. Samani, M., Sloman, M.: GEM: a generalized event monitoring language for distributed systems. Distributed Systems Engineering 4(2), 96–108 (1997)
26. Schmidt, D.C.: Model-driven engineering. IEEE Computer 39(2) (2006)

Using Diversity in Cloud-Based Deployment Environment to Avoid Intrusions

Anatoliy Gorbenko[1], Vyacheslav Kharchenko[1], Olga Tarasyuk[1],
and Alexander Romanovsky[2]

[1] Department of Computer Systems and Networks (503),
National Aerospace University, Kharkiv, Ukraine
{A.Gorbenko,O.Tarasyuk}@csac.khai.edu
V.Kharchenko@khai.edu
[2] School of Computing Science, Newcastle University, Newcastle upon Tyne, UK
Alexander.Romanovsky@ncl.ac.uk

Abstract. This paper puts forward a generic intrusion-avoidance architecture to be used for deploying web services on the cloud. The architecture, targeting the IaaS cloud providers, avoids intrusions by employing software diversity at various system levels and dynamically reconfiguring the cloud deployment environment. The paper studies intrusions caused by vulnerabilities of system software and discusses an approach allowing the system architects to decrease the risk of intrusions. This solution will also reduce the so-called system's *days-of-risk* which is calculated as a time period of an increased security risk between the time when a vulnerability is publicly disclosed to the time when a patch is available to fix it.

1 Introduction

Dependability is a system property which includes several attributes: availability, reliability, safety, security, etc [1]. In this paper we focus on ensuring security of web services by protecting them from malicious attacks that exploit vulnerabilities of system components. A computer system providing web services consists of hardware and a multitier system architecture playing the role of a deployment environment for the specific applications. The dependability of the deployment environment significantly affects the dependability of the services provided.

A web service application can be created as servlets and server pages, java beans, stored database procedures and triggers run in a specific deployment environment. This environment is constructed from a number of software components (Fig. 1). Typical examples of system components for web services are operating system (OS), web and application servers (AS and WS), and data base management systems (DBMS).

Vulnerabilities of operating system and system software represent threats to dependability and, in particular, to security, that are additional to faults, errors and failures traditionally dealt with by the dependability community [1, 2].

E.A. Troubitsyna (Ed.): SERENE 2011, LNCS 6968, pp. 145–155, 2011.

Intrusion tolerance is a general technique, which aims at tolerating system vulnerabilities that have been disclosed and can be exploited by an attacker. This is an active area of research and development with many useful solutions proposed (e.g. [3, 4, 5]). However, traditionally the focus has been on intrusion detection (e.g. [6, 7]), or combining different security methods [8, 9] (intrusion detection systems and firewalls, authentication and authorisation techniques, etc.). Less attention has been given to understanding how to make systems less vulnerable and to avoid intrusions while being configured and integrated out of COTS-components, such as OS, WS, AS, and DBMS.

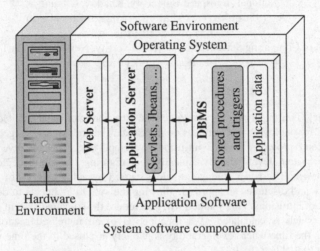

Fig. 1. General multitier architecture of a web system

In the paper we propose a general system architecture aiming at decreasing the risk of intrusion and reducing number of *days-of-risk* [10]. This architecture employs the diversity of the system software components and uses a dynamical reconfiguration strategy taking into account the current number of vulnerabilities (and their severity) of each component.

Such vulnerability information can be obtained by querying existing vulnerability databases (e.g. Common Vulnerabilities and Exposures, CVE and National Vulnerability Database, NVD), publicly available in the Internet. Besides, an implementation of the proposed architecture relies on the emerging cloud infrastructure services [11], known as Infrastructure as a Service (IaaS). IaaS provides a platform virtualization environment and APIs that can enable such dynamic reconfiguration by switching between pre-built images of the diverse deployment environments.

The rest of the paper is organised as follows. In Section 2 we describe a general architecture of intrusion-avoidance deployment environment making use of diversity of systems components and dynamical reconfiguration. Section 3 investigates diversity of the system software used as a deployment environment for modern Web applications. In Section 4 we discuss the vulnerabilities of some popular operating systems and demonstrate the feasibility of the approach proposed.

2 Intrusion-Avoidance Architecture

2.1 Diversity of Deployment Environment

Design diversity is one of the most efficient methods of providing software fault-tolerance [12]. In regard to multitier architecture of web-services, software diversity can be applied at the level of the operating system, web and application servers, data base management systems and, finally, for application software, both separately and in many various combinations. It is an obvious fact that some system software components may be incompatible with each other (i.e. Microsoft IIS and MS SQL can be run only on MS Windows OS series and incompatible with other operating systems). Hence, this fact should be also taken into account during the multiversion environment integration and selection of the particular system configuration.

Platform-independent Java technologies provide the crucial support for applying diversity of different system components. Thanks to JVM, Java applications can be run on different operating systems under control of various web and applications servers (see Table 1). These components form a flexible deployment environment running the same application software and allowing to be dynamically reconfigured by replacing one component by another one of the same functionality (e.g. GlassFish AS can be replaced with Oracle WebLogic, or IBM WebSphere, etc.). At the same time, the .NET applications can employ only restricted diversity of the deployment environment limited to Microsoft Windows series of operating systems and different versions of Internet Information Server and MS SQL.

Table 1. Diversity level and diversity components of Java deployment environment

Diversity Level	Diverse system components
Operating Systems (OS)	WinNT Series, MacOS X Server, Linux, FreeBSD, IBM AIX, Oracle Solaris, HP-UX, etc
Web-server (WS)	Apache httpd, Oracle iPlanet Web Server, IBM HTTP Server, lighttpd, nginx, Cherokee HTTP Server, etc
Application server (AS)	GlassFish, Geronimo, Oracle WebLogic, JBoss, Caucho Resin, IBM WebSphere, SAP NetWeaver, Apple WebObjects, etc
DBMS	MS SQL Server, MySQL, Oracle Database, Firebird, PostgreSQL, SAP SQL Anywhere, etc

2.2 Intrusion-Avoidance Architecture Making Use of System Components Diversity

The proposed intrusion avoidance approach is based on the idea of running at the different levels of the multitier system architecture (OS, WS, AS and DBMS) only those components having the least number of vulnerabilities. The rest diverse components should be hold in a stand-by mode.

When a new vulnerability is disclosed, the most vulnerable system component should be replaced with the diverse one having fewer numbers of open (i.e. yet unpatched) vulnerabilities. Such dynamic reconfiguration should also take into account severity and potential harmful consequences of different vulnerabilities, their popularity, availability of exploit code, etc. When a product vendor patches some vulnerability the system can be reconfigured again (after patch installation and re-estimation of the security risks). To compare the vulnerability level of the diverse components and spare configurations the weighted metrics can be used (1), (2). The proposed metrics estimating the system security risk can be also extended by taking into account other vulnerability attributes apart from severity and popularity.

$$VLC_i = \sum_{j=1}^{N_i} S_j \cdot P_j , \tag{1}$$

where VLC_i – vulnerability level (security risk) of the i-th system component; N_i – number of open (yet unpatched) vulnerabilities of the i-th component; S_j – severity of the j-th vulnerability; P_j – popularity of the j-th vulnerability.

$$VLS_k = \sum_{i=1}^{M_k} VLC_i , \tag{2}$$

where VLS_k – vulnerability level (security risk) of the k-th system configuration; M_k – number of system components (usually, each system configuration uses four basic system components: OS, WS, AS, DBMS); VLC_i – vulnerability level (security risk) of the i-th system component.

The general intrusion-avoidance architecture is presented in Fig. 2. The architecture employs the IaaS (Infrastructure as a Service) cloud technology [11] providing crucial support for dynamic reconfiguration, storage and maintenance of the images of spare diverse system configurations. The core part of such architecture is a configuration controller. It retrieves information about emerging vulnerabilities of the different software system components and information about patches and security advisories released by the companies (product owners). By analysing such information the configuration controller estimates current security risks, selects the less vulnerable system configuration and activates it. Other functions performed by the controller are patch and settings management of the active and spare diverse configuration.

3 Demonstration and Simulation

In this section we analyze vulnerability statistics of different operating systems gathered during 2010 year and show how the security risks would be reduced by employing the proposed intrusion-avoidance technique. To perform a demonstration we have developed the testbed software implementing the functionality of the configuration controller (see Fig. 3) and have simulated its decision-making and reconfiguration process by use of operating systems vulnerability statistics.

3.1 Vulnerability Databases

There are a number of databases (supported by commercial and government institutions) gathering and publicly providing information about software vulnerabilities: CVE (cve.mitre.org), NVD (nvd.nist.gov), XForce (xforce.iss.net),

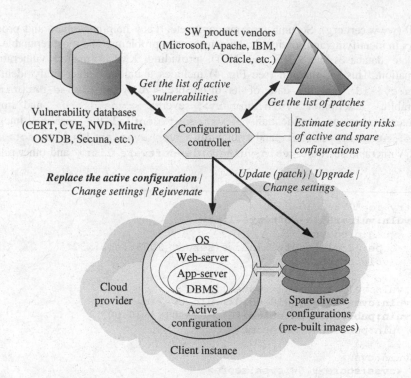

Fig. 2. General architecture of the cloud-based intrusion-avoidance deployment environment

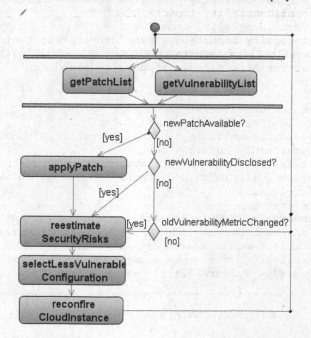

Fig. 3. Basic configuration controller's functionality: UML Activity diagram

CERT (www.cert.org), Secunia (secunia.com), etc. They help customers and product owners in identifying and solving the known security problems. The most reputable and complete databases are CVE and NVD, providing XML-formatted vulnerability information. This information (see Fig. 4) includes a unique vulnerability identifier `<vuln:cve-id>` and the date of its disclosure `<vuln:published-datetime>`, vulnerability description `<vuln:summary>`, severity `<cvss:score>` and impact attributes (impact on confidentiality `<cvss:confidentiality-impact>`, integrity `<cvss:integrity-impact>` and availability `<cvss:availability-impact>`), the list of vulnerable software `<vuln:vulnerable-software-list>` and other related data.

```
<entry id="CVE-2010-0020">
  <vuln:vulnerable-software-list>
    <vuln:product>
      cpe:/o:microsoft:windows_server_2008::sp2:x32
    </vuln:product>
    ...
  </vuln:vulnerable-software-list>
  <vuln:cve-id>CVE-2010-0020</vuln:cve-id>
  <vuln:published-datetime>2010-02-10T13:30:00
  </vuln:published-datetime>
  ...
  <vuln:cvss>
    <cvss:score>9.0</cvss:score>
    <cvss:access-vector>NETWORK</cvss:access-vector>
    <cvss:access-complexity>LOW</cvss:access-complexity>
    <cvss:confidentiality-impact>COMPLETE
    </cvss:confidentiality-impact>
    <cvss:integrity-impact>COMPLETE</cvss:integrity-impact>
    <cvss:availability-impact>COMPLETE</cvss:availability-impact>
    <cvss:source>http://nvd.nist.gov</cvss:source>
  </vuln:cvss>
  ...
  <vuln:cwe id="CWE-94" />
  <vuln:cwe id="CWE-20" />
  <vuln:references xml:lang="en" reference_type="UNKNOWN">
    <vuln:source>MS</vuln:source>
    <vuln:reference href="http://www.microsoft.com/technet/
      security/Bulletin/MS10-012.mspx" xml:lang="en">MS10-012
    </vuln:reference>
  </vuln:references>
  <vuln:summary>The SMB implementation in the Server service in
Microsoft Windows Server 2008 SP2 does not properly validate
request fields, which allows remote authenticated users to
execute arbitrary code via a malformed request, aka "SMB
Pathname Overflow Vulnerability." </vuln:summary>
</entry>
```

Fig. 4. Fragment of NVD's vulnerability record: an example

Unfortunately, none of existing vulnerability databases provides centralized information about patches and fixes available and exact dates of their issue. This complicates estimation of *days-of-risk* for each particular vulnerability. Assessment of the security risk and estimation of the number of open (i.e. yet unpatched) vulnerabilities of some particular software component can be done on the base of on-line monitoring of both vulnerability databases and vendor security bulletins.

To browse and query the NVD database we have developed a special software tool called *NVDView*. This tool allows to get an aggregated vulnerability information about the specified software product during the specified period of time. The results of vulnerability analysis of different operating systems obtained with the help of *NVDView* tool are discussed in the next section.

3.2 Operating Systems Vulnerability Analysis

In this section we provide a retrospective vulnerability statistical analysis of the different operating systems using information provided by NVD database. Table 2 reports a number of vulnerabilities disclosed during 2010 for Novel Linux v.11, RedHat Linux v.5, Apple MacOS Server v.10.5.8, Sun/Oracle Solaris v.10 and Microsoft Windows Server 2008.

Table 2. A number of vulnerabilities disclosed in 2010 for different operating systems

Operating System	Number of vulnerabilities disclosed per month												Total number
	Jan	Feb	Mar	Apr	May	Jun	Jul	Aug	Sep	Oct	Nov	Dec	
Novel Linux 11	3	7	5	11	5	4	0	1	33	5	18	23	**115**
RedHat Linux 5	3	7	5	11	6	4	0	1	32	4	18	23	**114**
Apple MacOS Server 10.5.8	2	0	25	0	0	10	0	3	0	0	18	0	**58**
Sun/Oracle Solaris 10	1	1	0	0	0	0	10	0	0	11	0	0	**23**
MS Windows Server 2008	3	16	8	14	2	5	2	14	5	6	0	16	**91**

It is clear, that the total (cumulative) number of vulnerabilities is not the best security metric to compare different software product. Since some period of time after vulnerability disclosure (called *days-of-risk* [10]) the product vendor issues a patch to fix the vulnerability. Thus, the most important characteristics is a number of the residual or open (i.e. yet unpatched) vulnerabilities. Cumulative and residual numbers of vulnerabilities for different operating systems starting from January, 1 2010 until December, 31 2010 is shown in Fig. 5 and Fig. 6. Unfortunately, none of the existing vulnerability databases provides the exact dates for patching the issues. As a result, it is not possible to estimate precisely how many vulnerabilities have remained unpatched by the specified date. Thus, in our work we used the following two basic assumptions:

1. We did not take into account number of residual vulnerabilities by the 1st of January 2010;

2. To eliminate fixed vulnerabilities in the Fig. 5 we used an average *days-of-risk* statistics provided in [13] (according to this survey, Microsoft Windows in average has 28.9 days-of risk; Novel Linux – 73.89 days-of risk; Red Hat Linux – 106.83 days-of risk; Apple Mac OS – 46.12 days-of risk and Sun Solaris – 167.72 days-of risk).

We believe that assumptions used do not affect our results as the purpose of this section is not to compare security of different operating systems but to demonstrate effectiveness and feasibility of the proposed intrusion-avoidance technique making use of operating systems diversity.

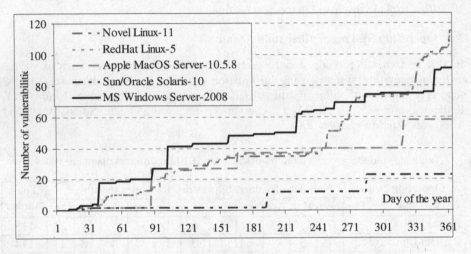

Fig. 5. Cumulative number of vulnerabilities in 2010 for different operating systems

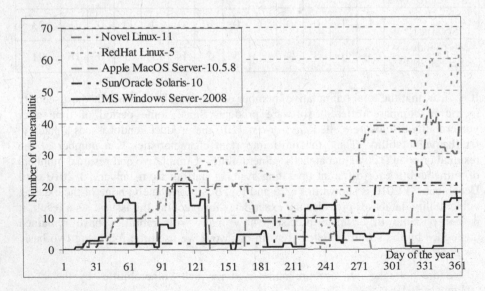

Fig. 6. Number of residual vulnerabilities in 2010 for different operating systems

3.3 Simulation of Dynamic Reconfiguration Strategy Making Use of Operating System Diversity

The results of dynamic operating system reconfigurations performed by the configuration controller taking into account the number of residual vulnerabilities (see Fig. 6) are summarized in Table 3. To simplify our demonstration we took out of consideration severity of different vulnerabilities.

In our simulation we used Novel Linux v.11 as the initial active operating system. Table 3 shows the set of subsequent switches between different operating systems in accordance with the vulnerability discovering and patch issuing process (see Fig. 5). The table also presents the exact dates and periods of active operation of different operating systems. In our simulation, the overall period of the active operation for Novel Linux v.11 was 33 days; for Apple MacOS Server v.10.5.8 – 152 days; for Sun/Oracle Solaris v.10 – 116 days and for MS Windows Server 2008 – 64 days.

Table 3. Operating systems reconfiguration summary

№	Active operating system	Operation period			Average number of open vulnerabilities
		start date	end date	duration	
1	Novel Linux v.11	01.01.2010	17.01.2010	17	0
2	Apple MacOS Server v.10.5.8	18.01.2010	18.01.2010	1	0
3	Novel Linux v.11	19.01.2010	25.01.2010	7	0.86
4	Sun/Oracle Solaris v.10	26.01.2010	06.03.2010	40	1.8
5	Apple MacOS Server v.10.5.8	07.03.2010	29.03.2010	23	0
6	MS Windows Server 2008	30.03.2010	30.03.2010	1	1
7	Sun/Oracle Solaris v.10	31.03.2010	14.05.2010	45	2
8	Apple MacOS Server v.10.5.8	15.05.2010	16.06.2010	33	0
9	Sun/Oracle Solaris v.10	17.06.2010	13.07.2010	27	1.48
10	MS Windows Server 2008	14.07.2010	01.08.2010	19	1.42
11	Apple MacOS Server-10.5.8	02.08.2010	24.08.2010	23	0
12	Novel Linux v.11	25.08.2010	02.09.2010	9	1.44
13	Apple MacOS Server v.10.5.8	03.09.2010	12.09.2010	10	3
14	MS Windows Server v.2008	13.09.2010	14.09.2010	2	2
15	Apple MacOS Server v.10.5.8	15.09.2010	15.11.2010	62	1.21
16	MS Windows Server 2008	16.11.2010	27.12.2010	42	4.86
17	Sun/Oracle Solaris v.10	28.12.2010	31.12.2010	4	11

As it can be seen from Table 4, the proposed approach to intrusion avoidance allows to hold the minimum possible number of residual vulnerabilities dynamically switching between diverse operating systems. It reduced the average *days-of-risk* to 11.21 and provided 146 vulnerability-free days.

Table 4. Intrusion avoidance summary

Operating system	Average days-of-risk [13]	Number of vulnerability-free days	Instant number of open vulnerabilities	
			max	avg
Novel Linux v.11	73.89	17	45	17.53
RedHat Linux v.5	106.83	17	63	23.05
Apple MacOS Server v.10.5.8	46.12	134	25	7.288
Sun/Oracle Solaris v.10	167.72	12	21	7.871
MS Windows Server 2008	28.9	27	21	6.466
Diverse intrusion-avoidance architecture with dynamic OS reconfiguration	**11.21**	**146**	**16**	**1.7**

During the remaining period of time the average instant number of the residual vulnerabilities would be equal to 1.7 that is almost four times as less as the best result achieved by Microsoft Windows Server 2008 (6.46 vulnerabilities at once).

4   Conclusions

The proposed intrusion-avoidance architecture that makes use of system component diversity can significantly improve the overall security of the computing environment used to deploy web services. Our work is in line with another recent study [14]. The approach proposed to intrusion avoidance is based on dynamical reconfiguration of the system by selecting and using the particular operating system, web and application servers and DBMS that have the minimal number of the residual (yet unpatched) vulnerabilities taking also into account their severity. Such strategy allows us to dynamically control (and to reduce) the number of residual vulnerabilities and their severity by the active and dynamic configuration of the deployment environment. This helps the architects to decrease the risks of malicious attacks and intrusions. The intrusion-avoidance architecture mainly relies on the cross-platform Java technologies and the IaaS cloud services providing the crucial support for diversity of the system components, their dynamic reconfiguration and maintenance of the spare configurations. The existing vulnerability databases like CVE and NVD provide the necessary up-to-date information for the security risk assessment, finding the least vulnerable configuration and reconfiguration decision making. The purpose of the paper is not to compare the security of various software components. Nevertheless, we would like to mention here that the least number of vulnerabilities in 2010 were disclosed in Sun/Oracle Solaris v.10, whereas the largest number of vulnerability-free days was provided by Apple MacOS Server v.10.5.8. At the same time, mainly due to the least *days-of-risk,* Microsoft ensured the least instant number of residual vulnerabilities in its Windows Server 2008. The vast majority of vulnerabilities of Novel Linux v.11 and RedHat Linux v.5 were vulnerabilities in the Linux core. Thus, they occurred in both operating systems and resulted in the similar curves of the cumulative number of vulnerabilities (Fig. 4). However, Novel spent more efforts on fixing security problems in its Linux distributive that was resulted in the lower *days-of-risk* (see Table 3).

Acknowledgments. Alexander Romanvosky is supported by the EPSRC/UK TrAmS platform grant.

References

1. Avizienis, A., Laprie, J.-C., Randell, B., Landwehr, C.: Basic Concepts and Taxonomy of Dependable and Secure Computing. IEEE Transactions on Dependable and Secure Computing 1(1), 11–33 (2004)
2. Cachin, C., Poritz, J.: Secure Intrusion Tolerant Replication on the Internet. In: Proc. International Conference on Dependable Systems and Networks, pp. 167–176 (2002)
3. Veríssimo, P., Neves, N.F., Correia, M.: The Middleware Architecture of MAFTIA: A Blueprint. In: Proc. 3rd IEEE Survivability Workshop (2000)
4. Pal, P., Rubel, P., Atighetchi, M., et al.: An Architecture for Adaptive Intrusion-Tolerant Applications. Special Issue of Software: Practice and Experience on Experiences with Auto-adaptive and Reconfigurable Systems 36(11-12), 1331–1354 (2006)
5. Nguyen, Q.L., Sood, A.: Realizing S-Reliability for Services via Recovery-driven Intrusion Tolerance Mechanism. In: Proc. 4th Workshop on Recent Advances in Intrusion-Tolerant Systems (2010)
6. Chatzis, N., Popescu-Zeletin, R.: Special Issue on Detection and Prevention of Attacks and Malware. Journal of Information Assurance and Security 4(3), 292–300 (2009)
7. Raggad, B.: A Risk-Driven Intrusion Detection and Response System. International Journal of Computer Science and Network Security 12 (2005)
8. Valdes, A., Almgren, M., Cheung, S., Deswarte, Y., Dutertre, B., Levy, J., Saïdi, H., Stavridou, V., Uribe, T.E.: An Architecture for an Adaptive Intrusion-Tolerant Server. In: Christianson, B., Crispo, B., Malcolm, J.A., Roe, M. (eds.) Security Protocols 2002. LNCS, vol. 2845, pp. 158–178. Springer, Heidelberg (2004)
9. Powell, D., Adelsbach, A., Randell, B., et al.: MAFTIA (Malicious- and Accidental-Fault Tolerance for Internet Applications). Proc. International Conference on Dependable Systems and Networks 35, 32–35 (2001)
10. Ford, R., Thompson, H.H., Casteran, F.: Role Comparison Report – Web Server Role. Security Innovation Inc., p. 37 (2005),
    http://www.microsoft.com/windowsserver/compare/
    ReportsDetails.mspx?recid=31
11. Buyya, R., Broberg, J., Goscinskin, A.M. (eds.): Cloud Computing Principles and Paradigms, p. 664. Wiley, Chichester (2011)
12. Strigini, L., Avizienis, A.: Software Fault-Tolerance and Design Diversity: Past Experience and Future Evolution. In: Proc. 4th Int. Conf on Computer Safety, Reliability and Security, pp. 167–172 (1985)
13. Jones, J.: Days-of-risk in 2006: Linux, Mac OS X, Solaris and Windows (2006),
    http://blogs.csoonline.com/days_of_risk_in_2006
14. Garcia, M., Bessani, A., Gashi, I., Neves, N., Obelheiro, R.: OS Diversity for Intrusion Tolerance: Myth or Reality? In: Garcia, M., Bessani, A., Gashi, I., Neves, N., Obelheiro, R. (eds.) Proc. Performance and Dependability Symposium at the International Conference on Dependable Systems and Networks, pp. 383–394 (2011)

'Known Secure Sensor Measurements' for Critical Infrastructure Systems: Detecting Falsification of System State

Miles McQueen and Annarita Giani

Idaho National Laboratory and University of California at Berkeley
Miles.McQueen@inl.gov, agiani@eecs.berkeley.edu

Abstract. This paper describes a first investigation on a low cost and low false alarm, reliable mechanism for detecting manipulation of critical physical processes and falsification of system state. We call this novel mechanism *Known Secure Sensor Measurements* (KSSM). The method moves beyond analysis of network traffic and host based state information, in fact it uses physical measurements of the process being controlled to detect falsification of state. KSSM is intended to be incorporated into the design of new, resilient, cost effective critical infrastructure control systems. It can also be included in incremental upgrades of already installed systems for enhanced resilience. KSSM is based on *known secure* physical measurements for assessing the likelihood of an attack and will demonstrate a practical approach to creating, transmitting, and using the known secure measurements for detection.

Keywords: Secure measurements, system state, control systems, cyber security, critical infrastructure.

1 Introduction

Software is a primary component of control systems which are used to operate our critical infrastructures such as the energy, water, chemical, transportation, and critical manufacturing sectors. To attain system resilience against disturbances it is critical to develop methods to enhance reliable state awareness. This discussion paper proposes a novel integrated hardware-software approach for quickly and effectively detecting adversarial manipulation of the physical infrastructure and the attacker's attempted falsification of system state. This detection mechanism is a step towards early response to disruptions. We present now a first investigation and the main concept and ideas. We plan to explore new automated adaptive responses based on this methodology.

The detection of network attacks has been studied for years. Many commercial Intrusion Detection Systems (IDSs) are used to provide defense in depth for IT systems. Due to the large amount of traffic in most IT systems, the issue of lowering the false positive rate without increasing the false negative rate cannot be overcome [1], [2]. This is a limit of current IDS systems.

E.A. Troubitsyna (Ed.): SERENE 2011, LNCS 6968, pp. 156–163, 2011.

A high false alarm rate is unacceptable for control systems due to the high costs required to respond to alerts. Discussions in the control system community focused on the idea that IDSs for control systems may have far fewer false alarms than that experienced by IT systems due to the greater regularity and reduced complexity of the messaging [3]. While work continues in this area, it is yet to be proven valid in anything other than artificial settings. Further, a continued focus on the network seems to ignore advantages that critical infrastructure control systems may provide for detecting intrusions.

One benefit is that control systems operate physical processes with several associated measurements. These measurements are used locally for real-time control and reported back to the operators for monitoring and control. If an adversary wishes to manipulate the process without early detection by an operator, the attacker must deceive the operator regarding the current system state. This would involve the modification of measured values being sent back to the operators or the injection of false measurements. If the measurements could be protected, the system state could be known by the operator with high confidence, permitting effective action. There are known cryptographic techniques for assuring the security of a message so that manipulation, injection of a new message, or replay of a previous message will be detected [5,4]. Unfortunately, many control systems have limited processor cycles available at the sensor for encrypting, sensor power may be limited, and in some cases the communication bandwidth will be in short supply. Consequently, a comprehensive solution for control systems which makes full use of standard cryptographic techniques is not generally practical.

We propose a low-cost technique that uses physical process measurements to build an intrusion detection engine faster and more accurate than what is currently feasible. Encrypting all measurements is not reasonable due to limited control system resources. Further, one cannot necessarily afford the time to encrypt before forwarding even a single message without negatively impacting the timely reception of the message at the controller. But the use of encryption is needed to provide enough known secure data to an engine to enhance attack detection. So the technique must be low cost, make limited demands on system resources, and must have flexible timing requirements.

2   Technical Objectives and Research Approach

Our objective is to investigate the value of KSSM for effective detection of unauthorized process manipulation and falsification of system state.

2.1   Targeted Facilities

We consider critical infrastructure control systems that currently lack robust cryptographic techniques, and have limited communication bandwidth and computational resources. Due to the long lifetimes of most control systems this description applies to most infrastructures. Thus, we anticipate a significant benefit from the deployment of the KSSM technique.

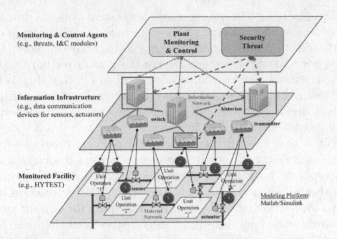

Fig. 1. Hybrid energy production facility

A hypothetical hybrid energy plant is shown in figure 1. This figure represents a hybrid energy production facility with three abstract layers. The lowest layer is the physical process which consists of a set of production units each of which consists of reactors, tanks, gas flows, coolers, heaters, valves and like physical components. The information layer, in the middle, is responsible for communication. The sensors in the physical layer communicate with control devices and commands are sent to the edge controllers that drive actuator behavior. The highest layer is represented by the primary functions of plant control, and security threat monitoring and alarming. These highest level functions make use of real time data feeds from the physical plant up through the communication layer, and may also make use of information derived over time through initial monitoring of the system (e.g. passive network discovery).

We assume that the attacker can compromise any of the components in the information layer without being detected as long as the attacker does not modify the sensor signals being transmitted back to the controller and the control room. KSSM is not designed to detect the system process exceeding its operational performance envelope, normal system monitoring is expected to detect that situation. Rather, KSSM is designed to reliably detect any attempt to falsify system state through manipulation of one or more of the sensor measurements being reported back to the control room.

2.2   KSSM System Hypotheses

We created the following four hypotheses to stay focused on the core issues in conceptualizing, designing, and validating a prototype of a KSSM system.

H1. A small set of sensor measurements, which are known to be secure, can significantly aid the operator and detection engine in more quickly and accurately identifying a cyber attack. A hybrid energy model, simulation, and detection engine will be used as an experimental foundation for measuring automated and human detection of process failures and falsification of system state when KSSM data is available.

H2. Some known secure measurements from randomly chosen sets of sensors providing data within selected time frames will harden the process against covert cyber attacks attempting to blind the operator and KSSM detection engine. Neither the enhanced operator effectiveness nor enhanced detection engine performance(gained from using a fixed set of known secure data) will be degraded by the changing and diverse sets of sensors selected for providing the known secure data.

H3. It is possible to create a very low cost, limited bandwidth, and highly secure measurement capture and communication channel for transmitting $k_i\%$ $(0 < k_i < 100)$ of a chosen sensor's physical measurements, end to end, from sensor to detection engine for analysis. The channel will involve adaptation of known cryptographic protocols to provide message and measurement integrity, and detection of replay attacks. Tradeoffs between cryptographic computational requirements at the sensor, power restrictions of a sensor, network bandwidth limitations, and the speed and accuracy of detection will be assessed in selecting specific cryptographic techniques for KSSM systems and establishing each KSSM sensor's appropriate value for k_i.

H4. Heuristics for selecting the set of sensors providing known secure sensor measurements can be developed which allow for the results of this research to be easily adapted for use in the design, implementation, and configuration of many diverse industrial control systems and infrastructures.

In a KSSM enabled infrastructure, the attacker will be unable to reliably falsify the process state to the control room operators.

2.3   KSSM Sensor

In any technique intended to protect against an intelligent adversary, one or more components must be trusted. We assume that the cryptographic sensor module, which includes the hardware or software which receives and encrypts the physical measurements and identifiers gained directly from the sensor hardware (e.g. AD converter), is trusted.

Figure 2 represents a KSSM hardware enabled sensor. The signal from the AD converter is tapped off and available to the secure encryption module. This module, at some randomized time δ after the unencrypted measurement M_i is sent, forwards the encrypted version E_i of the measurement value to the KSSM detection module running on a control room computer. Whether or not the encrypted version of the plaintext measurement is sent depends on whether that particular sensor is currently selected by the KSSM control room module, and whether the secure encryption module selects it as one of the k_i of measurements for which a dual encrypted value will be expected. We note that these sensor functions may also be implemented in software and reside in the sensor or closest

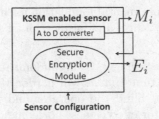

Fig. 2. Sample KSSM enabled sensor

computational edge point. The KSSM detection algorithm in the control room, which must also be trusted, will compare the two versions of the measured value, unencrypted and encrypted, and trigger an alarm if there is any difference. For the exposition of the idea in this paper, we are making simplifying assumptions related to reliable transport of measurements, both plaintext and ciphertext, and to the reliability of the sensor and encryption hardware.

3   KSSM and Attacker Scenarios

While not required, we assume that all process sensors are KSSM hardware or software enabled, all encryption modules are secure from cyber attack, and the KSSM control room module, including the detection engine, are secure. Any other component in the system, including the entire information infrastructure layer may be assumed to be compromised by an attacker.

3.1   Attack Scenarios

In this section we present different possibilities for the system and the attacker. Figure 3 presents two scenarios. The system on scenario 1 has no KSSM sensors available. The adversary has compromised choke points in the communication network and is behaving as a man in the middle by preventing all valid sensor signals to the control room and replacing them with corrupted, signals C_i. This deception could be done, as Stuxnet partially demonstrated, through collection and then replay of sensor measurement data. The attacker is now able to manipulate the process as desired while the operators remain completely unaware. This level of attack may be undetectable without KSSM and will leave the operators completely blind to the actual system state. KSSM sensors are available in scenario 2 and the attacker is choosing to corrupt only those signals which he knows are not providing encrypted values back to the detection engine. If the attacker corrupted the signals for which an encrypted version was sent at a later time then the corruption would be instantly recognized by the KSSM detector. Given that the encrypted signals are sent at some δ time after the unencrypted signal the attacker can only know which sensors will provide the encrypted signal by observing the network traffic for some period of time.

SCENARIO 1   **SCENARIO 2**

$M_i \rightarrow C_i$ $M_i \rightarrow C_i \vee M_i \vee (C_i \wedge E_i) \vee (M_i \wedge E_i)$ for all $i = 1$ to N.

Fig. 3. M_i is the measurement from sensor i. C_i is the corrupted version of measurement M_i. E_i is the encrypted version of measurement M_i. $\vee$ and $\wedge$ are the logic disjunction and conjunction.

SCENARIO 3

$$M_i \rightarrow C_i \vee M_i \vee (C_i \wedge E_i) \vee (M_i \wedge E_i) \vee (C_i \wedge E_i) \quad \text{for all } i = 1 \text{ to } N$$
$$\uparrow$$
$$\text{BLOCKED}$$

Fig. 4. Attacker identifies which sensors are providing encrypted versions of measurements. During the attack only a few of the sensors are blocked.

The attack in figure 4 consists in corrupting signals and blocking some of the encrypted versions. In fact attack would be easily detectable if the encrypted version of the measurement reached the detection engine and be compared against the corrupted version.

A way to make the above attack more difficult is to *periodically and unpredictably modify the subset of sensors providing encrypted values.* This ongoing and unpredictable selection of new sensors (and deselection of others) may be based on current system state, or communication network topology (for example let us not select our sensors such that their measurements all go through the same router).

3.2 KSSM Control Room Module

The KSSM module residing in the control room is represented in figure 5. It is responsible for modifying the subset of KSSM-enabled sensors which perform encryption, and is also responsible for detecting attacks. Many functions are needed to provide these capabilities and we will very briefly describe only the highest level functions.

The *system analyzer* receives input from network discovery tools which can both reside on the system and operate in real time, or can be one time only devices used during a phase such as system acceptance testing. It develops simplified models of the communication network to aid the sensor selection function in choosing smart subsets of sensors.

The *signal analyzer* is responsible for analyzing the sensor measurements that are provided to the control room, and alarming when appropriate. When encrypted and associated unencrypted values do not match then an alarm will be set; if some number of requested encrypted values do not arrive in a timely fashion, and are distributed over a variety of communication paths then it may

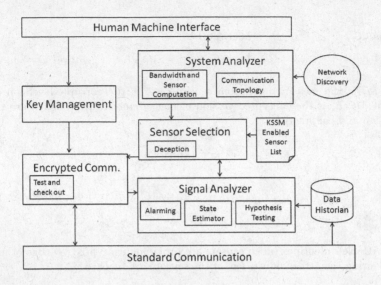

Fig. 5. Block diagram of KSSM module residing in the control room

be appropriate to raise an alarm based on probabilistic assessment of likely communication and sensor failures. Other conditions and analyses for alarming need to be explored.

The *sensor selection algorithm* will incorporate what is known about the communication topology and the failure rates of all components within the system. The failure rates may be based on empirical data or models built into the algorithm. Further some understanding of the limits on computation cycles available, sensor power restrictions, and limitations in communication bandwidth will be incorporated to aid, not only the selection of a new subset of sensors for KSSM but also the selection for each chosen sensor of the k_i of measurements that will be encrypted and forwarded. Sensor selection and the percent of measurements for which dual encrypted values are required may also be made selectable by the operators so that they have control in limiting the sensor processor cycles, sensor power consumption, and communication bandwidth utilized by KSSM.

The *cryptographic functions* will be adopted from currently well understood cryptographic components and systems. The KSSM-enabled sensor list is needed so that sensor selection can accommodate systems that are slowly being upgraded with KSSM-enabled sensors. And the KSSM user interface will be separate from all other devices in the control room in order to provide as much hardening against attack as possible.

4   Conclusion

The concept of Known Secure Sensor Measurements has been presented. The idea has been evaluated against potential attacker behavior and seems to have merit in providing early and reliable detection of attacker's possible attempts at falsification of process state. We are proceeding in our analysis and design of KSSM systems, and intend to initially validate through simulations in which effectiveness will be assessed through measuring the reduced time to detect falsification of system state. Eventually KSSM capability will be demonstrated on live control systems responsible for running our critical infrastructure facilities.

References

1. Axelsson, S.: The base-rate fallacy and the difficulty of intrusion detection. ACM Transactions on Information and System Security (TISSEC) 3(3), 186–205 (1981)
2. Fried, D.J., Graf, I., Haines, J.W., Kendall, K.R., Mcclung, D., Weber, D., Webster, S.E., Wyschogrod, D., Cunningham, R.K., Zissman, M.A.: Evaluating intrusion detection systems: The 1998 darpa off-line intrusion detection evaluation. In: Proceedings of the 2000 DARPA Information Survivability Conference and Exposition, pp. 12–26 (2000)
3. Linda, O., Vollmer, T., Manic, M.: Neural Network Based Intrusion Detection System for Critical Infrastructures. In: Proceedings of International Joint Conference on Neural Networks, pp. 1827–1834 (2009)
4. Stamp, M.: Information Security, 2nd edn., ch. 3-5, 9. John Wiley and Sons, Chichester (2011)
5. Ferguson, N., Schneier, B., Kohno, T.: Cryptography Engineering, ch. 3-7 (2010)

Author Index